Tropical Deforestation

Exploring Environmental Challenges
A Multidisciplinary Approach

SERIES EDITORS

Sharon L. Spray, Associate Professor of Political Science Elon University
Matthew D. Moran, Associate Professor of Biology Hendrix College

ABOUT THE SERIES

Exploring Environmental Challenges: A Multidisciplinary Approach is a series of short readers designed for introductory-level, interdisciplinary environmental sciences or environmental studies courses. Each reader, focused on a single, complex topic of environmental concern, outlines the concepts, methods, and current research approaches used in the study of that particular environmental challenge from six distinct fields of study in the natural sciences, social sciences, and humanities. This approach enables students and faculty alike to become familiar with a topic from perspectives outside their own training and to develop a broader appreciation of the breadth of efforts involved in investigating select, complex environmental issues.

TITLES IN SERIES

Global Climate Change, Sharon L. Spray and Karen L. McGlothlin
Loss of Biodiversity, Sharon L. Spray and Karen L. McGlothlin
Wetlands, Sharon L. Spray and Karen L. McGlothlin
Tropical Deforestation, Sharon L. Spray and Matthew D. Moran

Tropical Deforestation

EDITED BY
Sharon L. Spray and
Matthew D. Moran

ROWMAN & LITTLEFIELD PUBLISHERS, INC.
Lanham • Boulder • New York • Toronto • Oxford

ROWMAN & LITTLEFIELD PUBLISHERS, INC.

Published in the United States of America
by Rowman & Littlefield Publishers, Inc.
A wholly owned subsidiary of The Rowman & Littlefield Publishing Group, Inc.
4501 Forbes Boulevard, Suite 200, Lanham, Maryland 20706
www.rowmanlittlefield.com

PO Box 317
Oxford
OX2 9RU, UK

British Library Cataloguing in Publication Information Available

Library of Congress Cataloging-in-Publication Data

Tropical deforestation / edited by Sharon L. Spray and Matthew D. Moran.
p. cm. — (Exploring environmental challenges)
Includes bibliographical references and index.
ISBN-13: 978-0-7425-3481-0 (cloth : alk. paper)
ISBN 10: 0-7425-3481-2 (cloth : alk. paper)
ISBN-13: 978-0-7425-3482-7 (pbk. : alk. paper)
ISBN-10: 0-7425-3482-0 (pbk. : alk. paper)
1. Deforestation—Tropics 2. Forest biological diversity conservation—Tropics.
3. Forests and forestry—Tropics. I. Spray, Sharon L. II. Moran, Matthew D., 1969–
III. Series.

SD418.3.T76T75 2006
333.75'160913—dc22 2005031291

Printed in the United States of America

∞™ The paper used in this publication meets the minimum requirements of American National Standard for Information Sciences—Permanence of Paper for Printed Library Materials, ANSI/NISO Z39.48-1992.

Contents

About the Contributors vii
Preface xi
Introduction xv

1 DIVERSITY AND COMPLEXITY 1
A Biological Perspective on Tropical Forests
Matthew D. Moran

2 A CHANGING LANDSCAPE 25
A Geographical Perspective on Tropical Deforestation
Mark A. Cochrane

3 THE SWEET EARTH 49
A Biogeochemistry Perspective of Tropical Forest Soils
Deborah A. McGrath and C. Ken Smith

4 FROM FARMERS TO SATELLITES 77
A Human Geography Perspective on Tropical Deforestation
Peter Klepeis

5 TROPICAL TRADE-OFFS 103
An Economic Perspective on Tropical Forests
Erin O. Sills and Subhrendu K. Pattanayak

6 GLOBAL GOVERNANCE 129
An International Relations Perspective on Tropical Forests
Doris Fuchs

7 CONCLUDING THOUGHTS 155
Sharon L. Spray and Matthew D. Moran

Glossary *167*
Index *181*

About the Contributors

Mark A. Cochrane is an assistant professor at Michigan State University's Center for Global Change and Earth Observations. He has earned international recognition for his work documenting the severe effects of tropical forest fires that result from current human land use. Cochrane's interdisciplinary research draws upon ecology, remote sensing, and other fields of study in an effort to better understand the dynamics of land cover change. His Ph.D. is from Pennsylvania State University.

Doris Fuchs, professor of political science at the Leipzig Graduate School of Management (HHL) and the Wittenberg Center for Global Ethics, holds a Ph.D. in politics and economics from the Claremont Graduate University and has taught at the University of Michigan, Louisiana State University, and the Universities of Munich and Stuttgart prior to joining HHL. Her main fields of research are global governance, the role of business in politics, and sustainable development. Specifically, she explores the implications of the opportunities and constraints circumscribing the political roles of nonstate actors for the resolution of pressing societal problems.

Peter Klepeis, an assistant professor of geography at Colgate University in Hamilton, New York, received his Ph.D. from Clark University. His teaching includes courses on international environmental policy and the human dimensions of global environmental change as well as contributions to Colgate's interdisciplinary environmental studies program. His research explores historical and contemporary land use and land cover change in southeastern Mexico, Patagonia, and eastern Australia.

Deborah A. McGrath earned her Ph.D. in forest resources and conservation at the University of Florida, where she studied the biogeochemistry of forest conversion in Brazil's Amazon basin. She is currently an assistant professor of biology at the University of the South in Sewanee, Tennessee, where she teaches courses in ecosystem ecology, plant ecophysiology, and environmental studies and conducts research on land use change in the southeastern United States.

Matthew D. Moran received his B.A. in biology and Ph.D. in ecology from the University of Delaware, where he studied food web interactions in grassland communities. He currently is an associate professor at Hendrix College in Conway, Arkansas, where he teaches courses in ecology, field ecology, natural history, and zoology. He also organizes field experiences for undergraduates in Costa Rica, Ecuador, and the American Southwest, and he chairs the Environmental Studies Program. His research interests include grassland dynamics, fire ecology, and trophic interactions.

Subhrendu K. Pattanayak, Ph.D., fellow, and senior economist in environment, health, and development, specializes in nonmarket valuation, evaluation of ecosystem services provided by forest conservation, and economic analysis of environmental epidemiology. His research involves specifying testable hypotheses by applying economic theory to environment and development policies, conducting field experiments through household surveys in developing countries, matching the survey (microeconomic) data with meso-scale environmental and social statistics, and estimating econometric models to generate policy parameters and recommendations. Dr. Pattanayak is a research associate professor at North Carolina State University, an adjunct faculty member at Duke University, an associate editor for *Forest Science*, and a technical advisor for the South Asian Network for Development and Environmental Economists.

Erin O. Sills is associate professor and coordinator of international programs in the Department of Forestry and Environmental Resources at

North Carolina State University. She received her Ph.D. from Duke University and B.A. from Princeton University. Her research focuses on the economics of multiple-use forest management, including quantifying the value of nontimber benefits from forests, modeling the behavior of households who own or use forests, and evaluating the public benefits of forest policies and programs. This includes work on the nontimber benefits of tropical forests in Belize, Brazil, Costa Rica, India, and Indonesia. Dr. Sills is a vice president of the International Society of Tropical Foresters and a research associate of the Center for International Forestry Research.

C. Ken Smith earned his Ph.D. in forest resources and conservation at the University of Florida, where he studied plant-induced soil changes in the Brazilian Amazon. He currently chairs the Environmental Studies Program at the University of the South in Sewanee, Tennessee.

Sharon L. Spray, an associate professor of political science and environmental studies at Elon University in Elon, North Carolina, earned her Ph.D. from the Claremont Graduate University in Claremont, California. She teaches courses in American politics, international environmental policy, and domestic environmental politics and law. Her research includes work on presidential–congressional relations and domestic and international environmental policy formation.

Preface

In the last decade, a rapidly increasing number of institutions of higher education across the country have developed a wide variety of interdisciplinary programs in both environmental science and environmental studies. While many of these programs are centered primarily within the science curriculum, more and more institutions are strengthening their environmental sciences and environmental studies majors, minors, and concentrations by adding courses from both the social sciences and the humanities. The importance of integrating information from a variety of disciplines, including the sciences, social sciences, and humanities, has been recognized and considered in the design and revision of environmental curricula. Liberal arts institutions, in particular, are moving toward the development of inter- or multidisciplinary approaches as a basis for their environmental programs. These approaches are as varied as the institutions themselves. While many programs offer team-taught courses to provide true interdisciplinary approaches, others are built around a series of courses from across curricula that address environmental topics. The foundation for, and value of, such programs is the recognition that complex environmental challenges will necessarily require strengthening the interface between the social sciences, humanities, and natural sciences if we hope to find productive ways of addressing these issues.

To this end, many environmental programs across the nation are integrating innovative courses into their curricula that cross disciplinary boundaries. To what extent these courses are "multidisciplinary" versus "interdisciplinary" is often unclear. These two terms are frequently used interchangeably. For some, the distinction between the two may be of little consequence, but for others trying to identify texts that meet specific needs in the classroom, some clarification about this series may be in order.

By "multidisciplinary," we are referring to distinct disciplinary approaches to the study of a particular topic. Such perspectives do not preclude the integration of knowledge or material from other fields, but the interpretation of the information reflects a particular disciplinary perspective. We view this as a matter of disciplinary depth. As scholars, we necessarily cross the boundaries of knowledge and scholarship from other fields, but most of us have more depth in the field in which we received our academic training. Consequently, we interpret information through particular theoretical perspectives tied to our disciplinary training.

We view interdisciplinary teaching as the attempt at balanced integration of material from multiple disciplines. This, however, is a difficult goal when studying environmental issues. Most texts written about specific environmental issues reflect heavy bias toward the natural sciences, with some discussion of policy and economics, or, alternatively, the focus may be in the opposite direction, with an emphasis on policy and economics and limited discussion of science. More problematic is that many of the available texts fail to incorporate in any meaningful way the work of humanists, anthropologists, or sociologists—areas that we believe are essential for understanding complex environmental challenges.

This series was developed to facilitate interdisciplinary teaching in environmental studies programs by acknowledging that different disciplines bring distinctly different perspectives to the table and that scholars trained in those fields are best suited to explain these perspectives. The texts in this series are designed to assist faculty trained in a traditional social science, natural science, or humanities field to venture into areas of research outside of their own training.

The texts are purposefully balanced with half of the chapter contributions from the natural sciences and an equal number of chapters contributed from scholars in the humanities or social sciences. Each chapter identifies important concepts and theoretical perspectives from a particular field, and each chapter includes a supplemental reading list to facilitate additional study. We envision these texts to be the foundation for introductory environmental studies courses that examine environmental topics from multiple perspectives, or other courses that seek an interdisciplinary focus for the study of environmental problems. Because we anticipate that students from a variety of majors, both science and nonscience, will use these texts, the chapters are designed to be understandable to those with little familiarity of the topic or the field about which it was written.

The series is not neutral in its basic premise. The various topics in the series were chosen because we believe that the topics addressed are environmental challenges that we want students to better understand and for which we hope they will work toward future solutions. Individual authors, however, were asked to provide objective presentations of information so that students and faculty members could form their own opinions on how these challenges should be addressed. We care deeply about the environment, and we hope that this series serves to stimulate students to take the earth's stewardship seriously and promote a better understanding of the complexity of some of the environmental challenges facing us in this new century.

Introduction

In the southwestern region of the Amazon basin, nestled against the eastern flank of the Andes Mountains, lies the Brazilian state of Rondônia. The natural vegetation of Rondônia is tropical rainforest, an area of incredible diversity, supporting numerous species of plants, birds, mammals, arthropods, and fish. By the 1960s, most of the state was still completely covered in undisturbed forest. At this time, Brazil began a program of settlement of the area through the building of a portion of the Transamazon highway. The road brought in thousands of settlers who began clearing land for farming and ranching. By 1978, 7,800 square kilometers had been deforested as the migration began in earnest. By 1988, the amount of land deforested had risen rapidly to 58,000 square kilometers, and by 1996, it was up to almost 80,000 square kilometers, an area equal to the size of Missouri, or about 25% of the total land area of Rondônia. What had been rainforest was replaced by coffee plantations and cattle ranches. The year 1997 brought an unusually strong El Niño event that resulted in drought conditions throughout the region. Fires started by ranchers to clear land quickly got out of control and burned more of the previously undisturbed forest.

The settlement of this forest did not just affect the wildlife and plant life of the area. Indigenous people were pushed out or died out because of disease. Rubber tappers who could potentially exploit the forest sustainably were also pushed out by the new wave of settlers. Over time, the settlers who had moved looking for land found that the soils could not support small-scale intensive agriculture. Soon larger corporate farms began to dominate the area. Cities grew and expanded, and the capital, Porto Velho, reached a population of a quarter million people, making it the third largest city in the Amazon region.

Today the state of Rondônia is a typical province in the developing world. Cities are growing in size, forest is still being felled, and industries are developing. There is both wealth and poverty, often in uncomfortably close quarters. A large proportion of the wildlife has disappeared, and many species have probably become extinct. Yet much still remains, and that surviving **biodiversity** is spurring conservation efforts. The loss of forest and the problems associated with it are now recognized by the government and common citizens, although solutions are still elusive. Rondônia has therefore become a poignant example of the biological, cultural, and social changes that are occurring in areas of tropical forest throughout the world.

The chapters in this book represent different viewpoints, including the social, biological, and cultural issues that deal with the problem of tropical **deforestation**. Chapter 1, "Diversity and Complexity: A Biological Perspective on Tropical Forests," explores the causes of high species diversity in tropical forest and addresses the process of deforestation and how this human impact is threatening diversity. The author argues that this destruction can be halted and tropical forest restored, although the challenges to this goal are extensive. Chapter 2, titled "A Changing Landscape: A Geographical Perspective on Tropical Deforestation," focuses on the conversion of tropical forests to agricultural land, the most important factor in forest loss, and also discusses other processes that affect land use, such as fire, logging, mining, and development. The author concludes the chapter with a case study of the developing national forest system in Brazil, which could promote conservation and sustainable resource extraction to be implemented in other tropical countries. Chapter 3, "The Sweet Earth: A Biogeochemistry Perspective on Tropical Forest Soils," examines the impact of tropical deforestation on soil patterns and processes. Humans are ultimately dependent on healthy soils, and tropical soils present significant challenges for sustainable use. However, with careful management, the authors argue that tropical soils can produce food over the long term and that forests can be conserved.

Social scientists writing for this volume highlight a number of additional variables important for understanding causality, consequences, and the future of tropical forest management. In chapter 4, "From Farmers to Satellites: A Human Geography Perspective on Tropical Deforestation," the author further debunks the myth that tropical deforestation is part of a linear chain of events. Social, political, and economic variables interact to produce a complex, multidimensional mix of forces that shape resource extraction throughout the tropics. These variables include shifts in government policies, fluctuating commodity prices, and population pressures. Direct and indirect economic factors are further discussed in chapter 5, "Tropical Tradeoffs: An Economic Perspective on Tropical Forests," where the authors introduce the idea that some level of deforestation is socially optimal. Determining appropriate levels of deforestation, however, is difficult given that the distribution of costs and benefits is not spread equally throughout society. Hence, the authors suggest, correctly pricing the goods and services provided by tropical forests is an integral component of any future market-driven policy approaches for managing tropical forest resources to the benefit of societies as a whole. Finally, in chapter 6, "Global Governance: An International Relations Perspective on Tropical Forests," the author addresses international responsibility for deforestation. The author reviews a series of failed efforts in the international community to develop binding international agreements for managing tropical forests. While these efforts have not been abandoned totally, the author contends that international cooperation is more likely to occur in the form of private-sector cooperative efforts. Such efforts, however, are fairly recent, and more empirical research is needed to assess their viability.

While all of the authors in this book discuss a complex network of ecological, social, and cultural variables that contribute to tropical deforestation, they also provide us with a sense that there is still promise for preserving many tropical forest regions. Deforestation is not a phenomenon that, once started, cannot be slowed or stopped. Understanding the interface of variables and the degree to which each is at work in specific regions is inherently important in developing appropriate, regionally specific policy solutions. To this end, we hope that after reading this book, readers will have a better understanding of these variables and will be better equipped to understand what is at stake and what work lies ahead for scientists and policymakers in addressing this important environmental challenge.

HAPTER

1

Diversity and Complexity

A BIOLOGICAL ERSPECTIVE ON TROPICAL FORESTS

Matthew D. Moran

Introduction

For those who live in the temperate regions of the world, the first walk in a tropical rainforest is an experience unlike any other. The initial feature one notices is the overwhelming degree of green in the landscape. Towering trees supported by massive buttressed roots form the canopy, while a seemingly endless number of subcanopy layers step their way down toward the forest floor. Plants such as bromeliads and orchids grow upon other plants, and vines drape over the branches of the trees. These plants are covered by smaller plants and algae, all seeming to create a situation where every space is occupied by multiple organisms. Among all this green are hints of animal life. Birds call, insects buzz, and monkeys howl, revealing their presence through sound but seldom through sight. Within all this life, death is ubiquitous. Decay infiltrates one's sense of smell. Dead leaves, wood, and animals appear to decompose before one's eyes as their elements are absorbed back into living organisms. Plants are covered with spines and thorns, and deadly toxins reside inside their leaves. The tropical rainforest is not a soothing place, and it is not a tranquil park. It is dark and foreboding, and one cannot help the feeling of entering the "green hell" described by early explorers. Yet, for all these disconcerting characteristics, it is intensely beautiful, and no place on Earth feels more alive. It is the ultimate in biological development and complexity on this planet. Unfortunately, tropical forests are being dismantled rapidly by human activity, and their wealth of species is at risk of imminent extinction.

The biological diversity of tropical forests is an amazing result of millions of years of evolutionary development. The cause of this high diversity is a hotly debated topic in ecology. The complexity of the interactions among rainforest organisms is very poorly understood. Even the diversity itself is poorly known, as most species have not even been described scientifically. So, in many ways, tropical forests remain an unknown part of our natural heritage. This makes their loss even more tragic and disturbing, in that we are destroying something, but we do not know what it is.

Tropical Forest Characteristics and Biodiversity

Tropical forests fall between 30° north and 30° south latitude. The climate is characterized by warm conditions throughout the year, but other climatic conditions vary greatly among locations. Although **tropical rainforests** (also referred to as "humid tropical forests" or "lowland tropical forests") are most commonly discussed, they are not the only forest types present in tropical areas. Other forests such as savannas, cloud forests, tropical dry forests, and mangroves are some common examples. In general, a tropical forest is characterized by warm conditions, with rainfall varying according to the specific type of forest and the season of the year. In rainforests, the plant communities are dominated by broad-leaved evergreen trees that tend to form multiple layers from the canopy to the ground (i.e., stratified). These trees are often covered with numerous **epiphytes**, such as orchids, bromeliads, ferns, mosses, and even cacti. Vines are ubiquitous, and some large varieties, known as **lianas** (large woody species), make up a significant component of the forest biomass. Indeed, the biomass of vines and epiphytes attached to the trees can be greater than the biomass of the trees themselves. All these varieties contribute to the incredible plant diversity found in tropical rainforest habitats, and these plants support innumerable animal species.

Tropical dry forests and **savannas** are common habitats that also tend to support high species diversity. These habitats are characterized by a distinct and prolonged dry season during which rainfall is quite scarce. Therefore, trees are often deciduous, and epiphytes are much less common, although overall species diversity remains high. The difference between a savanna and a tropical dry forest is subtle. Savannas have large amounts of grassland with more scattered trees (e.g., east African savanna), while tropical dry forests tend to have a closed canopy of trees (although less dense than a rainforest). Another interesting subtype of tropical forest is the **cloud forest**, those areas at mid to high elevations that are almost perpetually cloaked in mist and fog. It is here that epiphytes reach their highest density, and the gnarled trees draped in mosses, ferns, and orchids give this forest a truly otherworldly feel. Tropical dry forests and cloud forests are even more endangered than lowland rainforests, since they occupy smaller areas, have a more temperate climate, and are therefore more likely to be settled by humans. **Mangroves** are unique habitats in which several species of trees have adapted to high salt content and are found lining areas where land and marine habitats meet. Mangroves are extremely productive habitats that support a great variety of terrestrial and marine species.

Tropical forests cover 6% to 8% of the earth's surface, yet they are estimated to contain over 50% of the planet's species. The incredibly high species diversity found in tropical regions is a characteristic that has been recognized for centuries and is one of the most profound patterns of life on Earth. It is known as the **species richness gradient** and is evident for the great majority of **taxonomic groups** of organisms. For instance, a single tree in the Amazon may contain more species of ants than the entire British Isles (Wilson 2002). Similarly, 830 species of birds have been recorded in Costa Rica, more than in the United States and Canada combined, even though Costa Rica is only 0.34% the size of these two northern countries. Colombia has twice as many birds as the United States does, despite being only one-tenth the size. A single hectare of rainforest may contain over 200 species of trees, much higher than the dozen or so that can be found in a typical hectare of temperate forest. The Amazon River system contains over 1,500 species of fish, more than the entire Atlantic Ocean (Wilson 1992). Insects are poorly known, but one study identified 1,200 species of beetles from a sample of 20 trees in a Panama rainforest (Erwin 1982). The trend for higher species diversity in tropical rainforests compared to other **biomes** is strong for most taxa, including birds, plants, insects, fungi, fish, and mammals, and some taxonomic groups are basically restricted to tropical climates (e.g., palms).

Although the pattern of higher diversity in the tropics has been known for hundreds of years, the causes of this pattern are still poorly understood, and the topic remains very contentious in the ecological sciences. No fewer than 28 hypotheses have been postulated to explain why tropical regions are more diverse (Rohde 1992), but no single hypothesis explains the pattern on its own (see table 1.1). It is more likely that a multitude of interacting mechanisms are responsible for tropical diversity. So, although some of the popularly hypothesized mechanisms clearly influence the diversity of tropical forests, much yet needs to be understood about the relative importance of each mechanism.

Some researchers have postulated that the tropics are older than temperate and polar regions. For instance, during the **Pleistocene epoch** (1.8 million to 10,000 years ago), glaciers advanced over much of North America several times, completely eliminating many habitats. Tropical areas, al-

Table 1.1. Major hypotheses developed to explain the global species richness gradient		
HYPOTHESIS	MECHANISM	LIMITATIONS
1. Evolutionary Time	The tropics are older and have had more time for speciation to occur.	The tropics may not have been stable for long periods of time.
2. Productivity	Areas with more primary productivity can support more species.	Many highly productive areas are species poor and many unproductive areas have high species diversity.
3. Heterogeneity	Areas of more spatial heterogeneity have more potential niches and support speciation and specialization.	There is no evidence that tropical areas are more spatially heterogeneous.
4. Competition	Competition is more prevalent in the tropics, which promotes specialization and allows more species to coexist.	Competition is common in nature, but global gradient is not well studied.
5. Predation	Predation is more common in the tropics, which reduces population sizes of organisms, reduces competition, and allows more species to coexist.	Predation clearly allows coexistence in many species, but it is unclear if predation is more common in the tropics.

(*continued*)

Table 1.1. (*Continued*)		
HYPOTHESIS	MECHANISM	LIMITATIONS
6. Disturbance	Moderate levels of disturbance promote diversity by reducing competition but are not so high as to cause extinction.	It is unclear if disturbance is different in tropical regions compared to other parts of the world.
7. Climate Stability	Areas that have fewer short-term climate fluctuations have reduced extinction rates, which allows species to accumulate over time.	Diverse tropical areas often have fluctuating climates.
8. Generation Time	Smaller species in tropics and lack of dramatic seasons allow for more generations per unit time and, subsequently, faster evolution rates.	More research is needed.

though reduced in area and perhaps made somewhat drier, were spared most of this destructive activity. Additionally, tropical regions appear to be less affected by climate change in general than do polar and temperate regions and therefore represent more stable areas. The presumed stability of tropical **ecosystems** has thus allowed more time for **speciation** and **niche specialization** so that more organisms exist there today. Research supports the notion that older habitats have more species, although there is much doubt about the long-term stability of the tropics. For instance, during the last glacial period, there is evidence that much of the Amazon Basin's rainforest reverted to drier savanna, leaving only isolated pockets of moist forest (Brown and Lomolino 1998).

A second popular theory is the productivity-richness theory. It states that more productive areas can support more species. This makes intuitive sense, since more **primary productivity** can potentially support larger populations, perhaps reduce resource competition, and subsequently reduce extinction rates. Indeed, the high productivity of the tropics may be necessary to support the incredible diversity seen. However, it may not be the cause of the high diversity. Tropical forests do not have the highest productivity of any ecosystem. At least two ecosystems, temperate rainforests (e.g., coastal forests of the Pacific Northwest of Canada and the United States) and coastal salt marshes, exhibit higher primary productivity. Temperate rainforests, although diverse, do not match the diversity of tropical forests, and a salt marsh is a very species-poor community. Further, some low-productivity systems such as the chaparral (e.g., southern California) have very high species diversity. Further research supports an alternative hypothesis that, within a region, moderate levels of productivity support the highest species diversity (Rosenzweig 1995).

Disturbance is predicted to enhance diversity by reducing competition for resources and creating more diverse **microhabitats**. Tropical forests appear to have a moderate level of disturbance through gaps formed by tree falls. Since the soil of many tropical forests is relatively poor, trees often have diffuse shallow roots that are adapted for nutrient uptake. However, these roots do not provide strong physical support. The weight added by epiphytes and liana vines further reduces the stability of trees so that storms often topple large ones, creating gaps in the rainforest. When seen from the air, one of the striking features of tropical forests is the presence of numerous gaps where large trees have fallen. These gaps produce different habitats that allow more species to coexist, and the relatively high mortality of falling trees prevents any one species from becoming dominant. Theory predicts that moderate levels of disturbance, high enough to reduce competition but not so high as to cause extinction, promote species coexistence, and this is supported by experimental evidence. Therefore this is a popular model, though there is still debate about the levels of disturbance in tropical forests compared to temperate and polar habitats.

Others have postulated that the relative short-term climatic stability of tropical regions promotes diversity. The daily and seasonal fluctuation in temperature and rainfall is much less extreme in many tropical regions (although still more so than is generally appreciated) compared to temperate and polar regions. There are also good **correlational data** showing the link between seasonal climatic constancy and species diversity of some taxa, so this hypothesis has some support as well.

There have been other hypotheses that predict differences in predation, competition, spatial heterogeneity, pollination rates, and generation times in tropical regions, all of which may contribute to rich diversity. It is clear that most of these hypotheses have validity. The main question remaining is the degree of importance for each process. It is likely that interactions of these processes contribute to the pattern. Processes may vary over geographic areas as well, further contributing to the complexity of tropical forest function.

Another important question is how the incredible number of species found in tropical forests coexist, especially considering that many species are relatively rare and often overlap greatly in niche characteristics. Unified neutral theory suggests that many species have similar competitive abilities so that **competitive exclusion** is rare, and success is determined more by random processes than by deterministic effects (Hubbell 2001).

The Loss and Degradation of Tropical Forests

THE EXTENT OF LOSS

To demonstrate the causes and consequences of tropical forest loss, the focus will be on tropical rainforests (i.e., humid tropical forests). The majority of intact tropical rainforests are found in three major bioregions: the Amazon Basin, equatorial west and central Africa, and Southeast Asia.

Smaller areas include much of Central America, Madagascar, northeastern Australia, the Caribbean Islands, and many parts of the Polynesian Islands. Although tropical rainforest areas are found across much of the equatorial portions of the globe, a few countries account for a majority of surviving forest. Brazil, Indonesia, and the Democratic Republic of the Congo have a very large proportion of what tropical forest remains today. The actions and policies of these three countries will determine if large continuous areas of tropical forest will survive into the future.

Deforestation is the removal of forest and the subsequent conversion of the land to other uses. Deforestation is the primary threat to tropical rainforests and to all the species that inhabit them. Determining the rate of deforestation is not easy, but recent advances in satellite capability have improved estimates greatly. In 1990, there were an estimated 1.15 billion hectares of tropical rainforest in the combined areas of Latin American, Africa, and Southeast Asia, accounting for most tropical forest areas in the world. By 1997, 1.116 billion hectares remained, which equals a 0.52% loss per year (equivalent to 5.8 million acres per year deforested; Achard et al. 2002). While this calculation takes into account only forest totally converted to other uses such as crops or pasture, further loss has occurred through degradation from logging and other activities. It is also important to realize that by 1990, much forest had already been lost or degraded. Although the actual amount is hard to determine, it is generally accepted that approximately 50% was lost before 1990 (Terborgh 1992), a loss that undoubtedly resulted in many species becoming extinct. If the current rate of deforestation continues, essentially no tropical forest will remain in 100 years.

Other types of tropical forest, such as savannas and dry forests, have also suffered great losses. Some areas, such as mangroves, are more endangered than tropical rainforests, often because of their smaller area and their greater risk of development (e.g., being located in coastal regions).

It is important to note that virtually no tropical forest that remains today is "pristine," or untouched by human activity. Indigenous people have occupied virtually all areas of tropical forest for thousands of years and have undoubtedly altered these ecosystems in profound ways. For instance, ecologists have often described the "silent forest syndrome" in areas occupied by large numbers of indigenous people in Amazonia. Although the forest appears normal at first inspection, a closer look shows that large mammals and birds are greatly reduced or absent. Well-regulated protected areas where hunting is prohibited often show notably higher mammal and bird populations compared to unprotected areas (Terborgh 1999). Many of these animals are probably important for pollination, population control of certain organisms, and seed dispersal. However significant the impact of indigenous populations on wildlife and species diversity, it is minimal compared to what is occurring today as modern technology invades tropical forests. Currently, forests are being eliminated completely, and entire ecosystems are at risk of extinction.

THE SPECIES–AREA RELATIONSHIP

In the face of the ongoing tropical forest loss, ecologists are attempting to determine what exactly is being lost. The **species–area relationship** (often called the first law of ecology) is one of the fundamental ways to explain species diversity on Earth, and it is an important way to estimate how many species are being lost in any degraded habitat, including tropical forests. The relationship shows that as the area of habitat increases, the number of species present increases. There are multiple reasons for this relationship. First, larger areas contain more habitat types (and by inference, more potential **niches**) and therefore more species. Larger areas also support larger population sizes of individual species, thus reducing the probability of extinction. Studies of island habitats as well as mainland ecosystems have verified that this is a consistent pattern across the globe.

The consistency of this relationship allows ecologists to estimate how many species will go extinct as habitat is reduced by human activity. Importantly, it does not require that we know how many species actually exist (which we do not in most tropical areas); we only need to know the area of the habitat lost. The general rule that has been stated from the species–area relationship is that a 90% reduction in habitat area will result in 50% of species being driven to extinction. For example, the Atlantic Forest, a thin strip of coastal rainforest in southeast Brazil noted for many **endemic** species, has already been reduced by almost 95%. Therefore, the prediction is that at least one-half of species unique to that area have already gone extinct. However, that is an oversimplification, as the 90-50 rule is not consistent, and different habitats have different extinction values. For instance, the range of species and the types of habitat destruction affect how many species will be lost in a particular ecosystem (Ney-Nifle and Mangel 2000). Regardless of the actual value, loss of habitat clearly causes species to go extinct.

The Atlantic Forest example is not atypical, though. As with most tropical ecosystems, the destruction occurred before the area was well inventoried, so we have little knowledge of what was lost, and therefore almost no knowledge of the characteristics that could help us predict actual species loss. From what remains, it is clear that the Atlantic Forest was very diverse, perhaps more so than the typical rainforest, as **endemism** rates are particularly high. The intensive management and captive breeding of some rare charismatic animals (e.g., the golden lion tamarin) has saved some of this habitat's unique biota from extinction. However, the species loss that had undoubtedly already occurred is a tragedy to science and to humanity. Similar patterns have occurred in other tropical areas, including the biological hotspots of Madagascar and Hawaii, once incredibly diverse tropical areas, now decimated by deforestation and other types of habitat alteration.

FRAGMENTATION

As the amount of tropical forest being lost makes the prospect for the future of many species seem bleak, even the surviving forest may not be of the same quality as large areas of unbroken forest. When forests are cleared, be they tropical or temperate, it is seldom done in a systematic way. Forests are cleared in a haphazard pattern, resulting in what ecologists call **fragmentation**. The resulting fragments of forest are often of varying size and shape, and these patterns can have profound effects on species diversity.

One of the longest-running studies of the effects of fragmentation has been performed on Barro Colorado Island, Panama. The island (15 km^2) was created during the construction of the Panama Canal, with the creation of Lake Gatun (circa 1913) used as a water source for the canal locks. Before construction, Barro Colorado was a hilltop surrounded by extensive areas of forest. After canal construction, it was established as a nature reserve and has remained well protected ever since. Beginning in the 1920s, the island underwent intensive ecological study, and much has been learned about tropical forest ecology (Brown and Lomolino 1998). One of the important findings has been the profound changes that occur when a tropical forest habitat becomes isolated. As expected, large animals that require large territories, such as jaguars and pumas, quickly became extinct. Subsequently, small herbivorous mammals increased greatly, thus increasing seed predation rates, which in turn is threatening the survival of many plants. Many other animals also disappeared. For reasons unknown, 25 species of birds that were once common on the island are now absent (Karr 1982). So what have we learned from studying Barro Colorado? One thing is clear from the studies performed on this island. Small fragmented patches of tropical forest will still support rainforest plants and animals, but these fragments will be drastically different from unfragmented areas. However, Barro Colorado Island is a single data point and is therefore of limited value in understanding the large-scale fragmentation effects that are occurring across the tropics.

A larger fragmentation experiment is being conducted in the Amazon basin under the guidance of ecologist Thomas Lovejoy. Brazil requires that landowners can clear only one-half of rainforest habitat while leaving the other one-half undisturbed (although this law is often violated). Dr. Lovejoy seized this opportunity to conduct one of the largest ecological experiments ever undertaken. First, unfragmented rainforest on the front lines of settlement was carefully surveyed to catalog all the species present.

Landowners were then persuaded to vary the size of the rainforest areas left undisturbed. Land areas varied from 1 to 1,000 hectares, and the surviving fragments were then followed over time to monitor their response. The loss of species from small fragments was pervasive. For instance, army ant colonies died out in smaller blocks of habitat. This caused antbirds, specialists that feed on arthropods fleeing army ant columns, to disappear as well. Collared peccaries (pig relatives), important herbivores and **ecosystem engineers**, also disappeared. Peccary wallows, which normally become small wetlands over time, then disappeared as well, causing many amphibian species to vanish. Most disturbing was the slow decline in the quality of the habitat that remained. Small blocks of habitat had much **edge habitat** that abutted the cleared areas. This edge habitat experienced higher wind speeds, higher temperatures, and reductions in humidity. Many plants and animals that could not tolerate these conditions disappeared. Unfortunately, the edge habitat appeared to have the tendency to move inward over time, slowly reducing the areas of high-quality forest. Today, fragments of tropical rainforest support many rainforest species, but they are very different in community structure compared to a large area of unbroken habitat (Brown and Lomolino 1998). Regardless of the pattern of fragmentation, the amount of interior habitat declines more rapidly so that eventually only edge habitat remains. Therefore, the amount of habitat remaining may not be the best measure of conservation success, and instead the quality of habitat needs to be considered.

Fragmentation can have other negative effects on organisms, increasing their risk of extinction, and these effects often take considerable periods of time to become evident. For organisms that have poor dispersal, populations that survive in small patches will undergo **inbreeding.** Inbreeding often causes the expression of deleterious genes, which can reduce the viability of a population over time, eventually leading to local extinction. Fragmentation also creates smaller populations of organisms. Small populations can more easily experience local extinction as random fluctuations drive these populations to zero.

Since fragments contain a higher proportion of edge habitat, species that prefer edge habitats become more common and can displace interior species. It is common to find interior species attempting to use small patches of habitat, but they often suffer higher mortality near the edges. This pattern has been seen repeatedly in temperate habitats. For instance, in many small patches of temperate forest, **neotropical migrant** birds often suffer high nest predation by edge predators such as crows, raccoons, and blue jays, as well as nest parasitism by brown-headed cowbirds (Brittingham and Temple 1983). It is likely that similar processes occur in fragmented tropical forests.

HOW AND WHY TROPICAL FORESTS ARE BEING LOST

The reasons for deforestation are complex, but the patterns of loss are quite consistent. Typically, logging roads are built into wilderness areas, and valuable trees (such as mahogany) are removed. The loggers often supplement their diet through the hunting of local animals (known as bush meat), resulting in significant declines in these species. Though this is damaging, what follows next is the major threat to tropical forests. After the logging companies leave, the land is often sold off to landless peasants, people who are afflicted by grinding poverty and who completely clear the land to grow crops. Usually, the cut timber is burned to release the nutrients from the living tissue into the soil. Crops can usually be grown for several years, but the incessant rains and poor soil soon contribute to crop failure. The settlers then move on to the next area of forest, and the cycle begins again (Wilson 2002). It is in this way that the forest is slowly consumed at the edges, a few acres at a time. Over the years, this has been the major cause of tropical forest loss, and it continues today.

Oil, gas, and mineral development also have proved damaging to many tropical forests. For instance, many gold deposits have been discovered in the Amazon Basin, resulting in large mines being opened. These mines typically have poor environmental regulations and severely pollute local areas, most commonly with the mercury used to separate gold from the surrounding sediment. Mercury **bioaccumulates** and is a great risk to many animals high on the food chain, including humans. Oil and gas development has further eroded many relatively pristine areas. Large oil deposits have been discovered in many tropical countries, particularly Ecuador, Venezuela, Peru, and the Congo, all countries with extensive tropical forests remaining. Ecuador is a good example of a country with significant areas of tropical forest and significant oil reserves within those rainforests. The resulting oil operations in Ecuador have resulted in massive pollution, displacement of indigenous people, destruction of important wildlife areas, and extensive settlement in oil exploration areas. Even supposedly protected areas (e.g., Yasuní National Park, a **biodiversity** hotspot and possible **Pleistocene refugium**) have been infiltrated by oil companies with predictably negative consequences for the wildlife.

Why societies allow this unproductive and destructive process to occur is a difficult and multifaceted question. Most tropical forest is located in countries with high poverty rates. Since basic needs are not being met, nature preservation is not a priority for the government or the human population as a whole. Unfortunately, there is little evidence that clearing tropical forests aids in economic and social development, and by providing people with unproductive lands, it may actually contribute to social problems.

The root cause of tropical deforestation, and the root of all environmental problems, is the population growth of the human species. In the last 100 years, the human population has risen from 1 billion to approximately 6.2 billion. With this rise in population has come the concurrent increase in farms, roads, cities, houses, and economic activity. As populations swelled, the desire for land forced the population into sparsely populated, often marginal areas. The huge human population of today requires an immense quantity of food, and much tropical forest has been cleared for this required agricultural production. Today, the human population continues to grow at a global rate of about 1.4% per year, equal to an increase of 85 million people per year. Much of this growth is occurring in countries that contain most of the remaining tropical forest. Ultimately, without a stabilization of the human population and a gradual decline to a more sustainable level, it is unlikely that tropical forest ecosystems will survive into the future.

Global climate change has been implicated in the loss of some tropical forests, although the link between global warming and forest damage has been difficult to establish with certainty. There is evidence that global warming is causing an increase in the frequency and severity of El Niño events. El Niño, a periodic warming of the eastern equatorial Pacific Ocean, has dramatic effects on weather patterns across the world. During strong El Niño events, many tropical forest areas, such as the Amazon Basin and Indonesian islands, show severe declines in rainfall. During the 1998–1999 El Niño, these areas subsequently suffered large wildfires that consumed millions of acres of rainforest, much of it relatively undisturbed. Many well-known rainforest animals, such as the Orangutans of Borneo, exhibited severe population declines. Although most fires were started by humans, the unusual climate patterns allowed them to burn out of control.

Relatively minor changes in climate could also have adverse effects on many tropical forests. In the famed cloud forests of Monteverde in Costa Rica, the clouds and mist that typically hover above the forest have been rising in elevation. Many areas on the lower slopes now receive substantially less moisture than in previous years. Climatologists have suggested that this change in cloud level is caused by the gradual warming that has occurred during the last 100 years, warming attributed to human activity. Monteverde has also suffered declines and extinctions of several amphibian populations, including unusual species such as the Golden Toad (*Bufo periglenes*). It is likely that these two events are related, and this is a cause of great concern for species in other tropical areas.

All the mechanisms described above show few signs of abatement, indicating that tropical forest decline and subsequent species extinction will continue. The question remains as to how much tropical forest and how many species will survive. Confounding this question is the degree to which tropical forests can recover from the onslaught that is occurring now.

REFORESTATION

As Europeans settled North America, forests were cleared for agriculture and settlement in a pattern remarkably similar to what is happening in tropical countries today. In the year 1492, approximately 99% of the land east of the Appalachian Mountains in the modern-day United States was forested. Over the next 300 years, practically all of it was cleared and farmed. However, if one examines the eastern United States today, it would be hard to believe that this event happened, as much of the land is cloaked in rather impressive temperate forest. As settlers moved westward, they found the rich soils and prime agricultural areas of the Midwest and Great Plains. Production was higher in these areas, and eastern farmers could not compete. The result was an abandonment of farms in the east, and the land was allowed to go fallow. What followed was one of the greatest reforestation processes ever seen. Millions of acres underwent **succession**, and today, over two-thirds of the area is covered in forest. This forest, although not identical to the presettlement ecosystem, is rich in species and supports healthy populations of many animals. Even large mammals, such as the Black Bear (*Ursus americanus*), have recolonized much of their former range.

The question then emerges: Will tropical countries follow this model of settlement patterns, and are we witnessing a temporary tropical forest loss that can be reversed over time? Will we see species that are now declining increase and recolonize their former ranges, as has happened in North America, or are the tropical losses permanent? Current ecological research indicates that this is a complex question, and the answer may not be the same in all parts of the tropical world.

There are many reasons to believe that tropical forests will not recover from massive clearing and naturally regenerate as have the temperate forests in North America. First and foremost, tropical forests maintain most of their nutrients in the living tissue, resulting in soils of low fertility (see chapter 3). Normally, when a tree dies in a tropical forest, it is broken down quickly by the myriad bacteria, fungi, and insects, and the nutrients rapidly return to other living things. Very little of the organic matter and nutrients therefore ends up in the soil. The result is that there is a tremendous amount of biological activity being supported on a very poor substrate. When the forest is cleared over a large area and crops are grown and harvested, the soil quickly loses the ability to support growth, and it must be abandoned. Many areas of previously cleared and abandoned rainforest have not returned to forest after many years of opportunity, and have instead reverted to poor-quality and easily eroded grasslands. Madagascar is a particularly good example of this process, and it suffers from some of the world's severest erosion in areas formerly covered in rainforest.

Another major issue is the preservation of species during deforestation. Tropical organisms, be they plants or animals, tend to have smaller populations and occupy smaller ranges. The Black Bears of North America were able to reclaim much of their former territory because they still survived in smaller parts of their once extensive range. The same holds true for most temperate species that tend to exist over a large area. Tropical species, by contrast, are often highly localized. Therefore, it is possible to eliminate many species' entire habitat just by clearing a small area of forest. In a famous example of barely known, then lost diversity, Alwyn Gentry and Calaway Dodson discovered a botanically rich ridge of forest in Ecuador in 1978. This ridge contained many new species, including understory plants that had black leaves, a unique characteristic in the plant kingdom. This ridge was settled by people soon after it was biologically inventoried, which resulted in the total clearance of all native vegetation, including all the black-leaved plants. Many of the species discovered there have never been seen anywhere else and are now presumed extinct (Wilson 1992), a tremendous botanical loss. Assuming this pattern is similar across many tropical regions, if large areas of tropical forest regenerate, the restored forest may support only a fraction of its original diversity.

The third concern is that tropical forests have such a profound effect on climate that they may in part control their own weather, and by extension their own existence. For instance, approximately 50% of the rainfall in the

Amazon basin originates from **evapotranspiration** through the plants instead of evaporation from bodies of water (Salati and Vose 1984). One can often see fog and mist rising above a tropical forest, moisture that will soon form the rainfall that maintains the area as a tropical wet forest. Eventually, if enough forest is cleared, it is possible that rainfall will diminish to the point where tropical wet forests can no longer be supported, and reforestation will be impossible.

As bleak as the above evidence appears, there is support for the idea that tropical forests do have great regenerative ability, and the prevailing public notion that the clearing of tropical forest is always permanent is false. Pre-Colombian history gives us some profound insights into the possibility of natural reforestation. The Mayan Empire once occupied large portions of Central America, including most of Belize, Guatemala, Honduras, and the Yucatán Peninsula of Mexico. The Mayans built impressive cities and sustained large populations by utilizing advanced agricultural systems. For instance, by 900 CE, it is estimated that the city of Tikal (in modern-day Guatemala) was inhabited by over 100,000 people. The area of forest cleared was undoubtedly extensive. In fact, there is evidence that the collapse of the Mayan civilization was due in part to loss of soil fertility under a regime of intensive agriculture. However, after the Mayan collapse and the subsequent reduction in human population size, tropical forests regenerated. The regeneration was so complete that many large Mayan cites remained undiscovered until recent times, as the cloak of forest kept them hidden from view. Even today, moderately sized Mayan structures are discovered frequently. The tropical forest of today that surrounds Tikal is very diverse and is basically indistinguishable, in terms of species composition, from areas always inhabited by low numbers of people, an astounding fact considering the amount of disturbance that the Mayans must have incurred.

Recent evidence also indicates that some large human populations existed in the Amazon basin, a place generally thought to have always contained small numbers of people. The settlements in this area were large, elaborate, and connected by a sophisticated system of roads (Heckenberger et al. 2003). Alteration of the forest must have been profound, yet today the forest in this area is diverse and complex, even though the civilizations existed until approximately 1600 CE. Therefore, at least some, if not most, tropical forest appears to have substantial regenerative abilities.

The conventional notion that all tropical soils are poor is also not true. For instance, much of Central America and western South America is covered by young, rich soils of volcanic origin. This soil tends to retain fertility after years of proper farming and appears to reforest relatively easily. It is common in these areas to see abandoned farms quickly reforest.

One of the classic examples of recent reforestation has been the work performed at Guanacaste National Park on the Pacific side of Costa Rica. Here, Daniel Janzen and other scientists have taken on the difficult task of reestablishing a tropical dry forest. Much of the land was bought for conservation purposes during the 1980s, fire was excluded from the park, and the forest then began to regenerate quickly. Today, there is a developing forest that looks quite similar to the original vegetation, and many of the characteristic animals have returned. Most remarkable has been how quickly the reforestation has occurred, in an area that had been grazed heavily for decades. Similarly, lowland rainforest on the Caribbean side of Costa Rica is currently undergoing extensive reforestation in areas where banana plantations have been abandoned. The reforestation is now to the point where a new national park has been proposed for the area.

Further evidence of the potential resiliency of tropical forests is the presumed reduction in tropical forest area in the Amazon basin during the Pleistocene glacial periods. The Amazon contains 6 to 10 areas of high endemism (i.e., biological hotspots). These areas are hypothesized to have been portions of rainforest (surrounded by grassland or savanna, see page 3) that existed during the dramatically drier Pleistocene glacial periods (Haffer 1969). If these areas truly were the only remnants of rainforest, then a natural reduction of approximately 80% had occurred and was maintained until 10,000 years ago, when the glaciers began to recede. Yet today, most of the Amazon basin is covered in luxuriant and highly diverse tropical rainforest once again. Although this would seem to indicate the great regenerative ability of tropical forests, the **refugia hypothesis** is controversial, and other hypotheses have been developed to explain the same patterns (Prance 1982).

So the question remains, does tropical forest regenerate or not? The answer is complex and depends upon location, intensity of past use, and amount of damage incurred. Many areas that have rich soils and have not had intensive agricultural use probably will regenerate reasonably well. Even heavily abused areas may reforest over longer periods of time and support high species diversity, assuming that enough nearby forest survives to provide a refugium. However, some areas, especially those that have old, highly weathered soils, appear to resist reforestation over considerable periods of time. This has been particularly prominent in parts of the lowland Amazon Basin that were cleared, farmed, and then abandoned. Even if regeneration occurs, it is not clear that these new forests will support the levels of diversity seen prior to disturbance, especially in locations that are noted for high rates of endemism. Given the small range and often small population sizes of many species, it is therefore likely that the massive forest clearing occurring today will result in significant species loss, even if much of the forest can be reestablished.

Conclusion

Slowing, halting, and eventually reversing tropical forest loss will not be an easy process. Population growth, poverty, corruption, and economic expansion all work against the goal of preserving tropical forests. However, some countries have begun to address the myriad issues that inhibit forest preservation and have found some level of success. The model most often cited is Costa Rica. Costa Rica is a small country in Central America noted for its political stability, its progressive government, and its relatively prosperous people. It is also a hotspot of biodiversity and, fortunately, a place where many natural systems still function. Costa Rica also has a unique government among Latin American countries. Following the overthrow of the military government in 1949, democracy was established, and the military was abolished. Progressive taxation and an emphasis on education and health care have transformed Costa Rica into a peaceful, moderately wealthy country. As standards of living rose, government leaders and the population as a whole began to embrace conservation, so that by the 1980s, a significant system of national parks, refuges, preserves, and private reserves began to evolve. Today, approximately 35% of the land area is preserved at some level. With the well-crafted promotion of ecotourism, Costa Rica is now known as the one place in the world where tropical forest conservation is actually working. Admittedly, there are still problems. Deforestation on private land is still significant, poaching remains a problem, and continued population growth threatens the gains that have been made. However, there is an overwhelming sense of accomplishment in Costa Rica, and one can see the day when a truly sustainable society is established. The question remains whether this model can be replicated in other countries. Other tropical countries such as Ecuador, Belize, Panama, and Thailand have made significant strides in forest protection, although the future of these programs is in doubt.

So what might the future hold for tropical biodiversity? It is likely that deforestation will continue unabated for years to come. The amount of primary forest will undoubtedly decline as settlement continues, and many tropical species will undoubtedly go extinct, many before ever being described. Climate change, poaching, pollution, and resource development will likely put additional pressure on tropical forests. If current trends continue, the tropical world will be a less species-rich place in the future. However, there is hope that this trend can be reversed. Human population growth is slowing in many of the countries that contain tropical forests. Ecotourism is becoming a viable economic activity for some regions. Tropical forests are demonstrating stronger regenerative abilities than once thought. Most importantly, humanity is beginning to recognize the beauty, value, and importance of the tremendous diversity of life found in tropical forests. The developing appreciation for this splendid community of life creates hope that tropical biodiversity in all its forms will survive and continue into the future.

REFERENCES

Achard, F., J. D. Eva, H. Stibig, P. Mayaux, J. Gallego, T. Richards, and J. Malingreau. 2002. Determination of deforestation rates of the world's humid tropical forests. *Science* 297:999–1002.

Brittingham, M. C., and S. A. Temple. 1983. Have cowbirds caused forest songbirds to decline? *BioScience* 33:31–35.

Brown, J. H., and M. V. Lomolino. 1998. *Biogeography.* Sunderland, Mass.: Sinauer Associates.

Erwin, T. L. 1982. Tropical forests: Their richness in Coleoptera and other arthropod species. *The Coleopterists Bulletin* 36:74–75.

Haffer, J. 1969. Speciation in Amazonian forest birds. *Science* 165:131–37.

Heckenberger, M. J., A. Kuikuro, U. T. Kuikuro, J. C. Russell, M. Schmidt, C. Fausto, and B. Franchetto. 2003. Amazonia 1492: Pristine forest or cultural parkland? *Science* 301:1710–13.

Hubbell, S. P. 2001. *The unified neutral theory of biodiversity and biogeography.* Princeton, N.J.: Princeton University Press.

Karr, J. R. 1982. Avian extinction on Barro Colorado Island, Panama: A reassessment. *American Naturalist* 119:220–39.

Ney-Nifle, M., and M. Mangel. 2000. Habitat loss and changes in the species-area relationship. *Conservation Biology* 14:893–98.

Prance, G. T. 1982. *The biological model of diversification in the tropics.* New York: Columbia University Press.

Rohde, K. 1992. Latitudinal gradients in species diversity: The search for the primary cause. *Oikos* 65:514–27.

Rosenzweig, M. L. 1995. *Species diversity in space and time.* New York: Cambridge University Press.

Salati, E., and P. B. Vose. 1984. Amazon basin: A system in equilibrium. *Science* 225:129–38.

Terborgh, J. 1992. *Diversity and the rainforest.* New York: Scientific American Library.

———. 1999. *Requiem for nature.* Washington, D.C.: Island Press.

Wilson, E. O. 1992. *The diversity of life.* Cambridge, Mass.: Harvard University Press.

———. 2002. *The future of life.* New York: Knopf.

SUGGESTED READINGS

Terborgh, J. 1999. *Requiem for nature.* Washington, D.C.: Island Press.

Wilson, E. O. 1992. *The diversity of life.* Cambridge, Mass.: Harvard University Press.

Wilson, E. O. 2002. *The future of life.* New York: Knopf.

CHAPTER

2

A Changing Landscape

A Geographical Perspective on Tropical Deforestation

Mark A. Cochrane

Introduction

Land cover and land use change is a commonly used phrase that provides a catchall description of how both the vegetation covering the land and the human uses of terrestrial portions of the globe vary through both time and space. Typically, the changes in land cover are presumed to relate either directly or indirectly to human activities. Furthermore, although left unstated, common usage of the phrase refers to relatively recent changes, occurring over a period of years, decades, or centuries. Implicit in this meaning are the interactions between humans and their environment as they result from or affect human land use choices and practices. In short, it is the study of how people affect and are affected by their surroundings. The science *per se* is highly interdisciplinary since there are many ways to look at human-environment interactions. Whether approaching the problem from a natural or social science perspective, one must always account for both the human and the environmental components. The investigation of land cover and land use change involves the combination of empirical observations, remote sensing, ecological understanding, and both spatial and temporal representations of the dynamic processes involved.

In the tropics, the main category of land cover and land use change study relates to **deforestation**, or, more accurately, the conversion of forests into agricultural lands. The general premise is that changes in the extent and intensity of human land use in the tropics are resulting in the observed rapid decline in the amount of forest cover. There is much concern about these changes due to the perceived importance of these forests to global **biogeochemical** cycles and the great **biodiversity** that they contain. This chapter will discuss deforestation, its relation to land use, and the changes in ecological processes that are occurring across the tropics. It will also explore some of the myths surrounding tropical deforestation and some recent efforts to restore an element of harmony between human populations and the forests upon which they depend.

The Tropics

The tropics are defined as the region of the globe between the Tropic of Cancer and the Tropic of Capricorn. This corresponds to approximately 2,600 km (1,600 miles) north and south of the equator and covers one-quarter of the earth's surface. This portion of the earth is generally characterized by limited seasonality, which is related more to changes in precipitation patterns than to changes in temperature. The average mean temperature in the tropics ranges between 23°C and 27°C. Intense sunshine is present for 12 hours a day the entire year and leads to rapid rates of evaporation. However, many areas of the tropics are also characterized by heavy rainfall that ranges between 1.5 to 4 or more meters per year.

Tropical rainforests cover much of the land area of Earth's equatorial region. Although **pantropical**, the world's tropical rainforests are divided into three major formations. By far the largest is the formation in the Americas. This zone comprises much of the northern half of South America east of the Andes Mountains and north of the Chaco (scrub and palm forests of Paraguay and Bolivia). It covers half of Brazil; the eastern portions of Bolivia, Ecuador, and Peru; and most of Colombia, Venezuela, Suriname, and the Guianas. The forests extend through much of Central America and up into southern Mexico, as well as into portions of many Caribbean island nations. The Amazon accounts for much of the area of this formation. The Atlantic rainforest along the eastern coast of Brazil is isolated from the rest of the formation.

The second largest tropical rainforest formation is found primarily in Southeast Asia. Largely existing north of the equator, it sprawls across many nations of southern Asia. The isolated forests of the Western Ghats in southern India and the forests of Sri Lanka bound the region to the west. It ranges north into southern China and the many islands of the Philippines, and it extends east out into the Pacific along the extensive island nations of Indonesia, New Guinea, and the Solomon Islands. To the south, it covers the northern fringes of Australia. It is noteworthy that the New Guinean and Australian forests are biogeographically distinct from the other forests (MacDonald 2003; Wallace 1876).

The smallest of the major tropical rainforest formations exists in Africa. It is largely restricted to a belt roughly 1,100 km in width centered on the equator, with the majority of the remaining forests in West Africa and a few relict patches in East Africa. This formation was previously much more extensive, and relict patches of tropical rainforest exist all of the way to the Indian Ocean, but they are much less extensive than the **tropical dry forests** and **savannas** that characterize the region.

Although this chapter will concentrate on the **evergreen tropical forests**, it should be understood that land cover in the tropics also includes so-called tropical dry forests, which are **deciduous** and which transition into **woodlands** (regions of sparse, widespread tree cover) and savannas (continuous strata of herbaceous plants, especially grasses and sedges). Although less publicized, human land use and the resultant land cover changes have made the tropical dry forests and **cerrado vegetation** (represented by a wide range of open woodland, open scrub, and highly species-diverse grassland forms) some of the planet's most threatened **ecosystems.**

Tropical Rainforests

Tropical rainforest is somewhat of a misnomer for many of the forests throughout the tropics, as many of these forests persist even in areas that are annually subjected to several months of dry season. The expression "dry season" is a bit misleading here, in that in the tropics it is usually characterized by a fair amount of rainfall (up to 150 mm/month). The distinction between wet and dry seasons is generally accepted to be the contrast between months that have sufficient rainfall to balance the **evapotranspiration** demands (100 to 150 mm/month) and those that do not. Therefore, during the dry season, the forest requires more moisture than is being provided by precipitation. In practice, this means that soil water is being used to make up the deficit. One of the myths of the tropics is that the forests are universally shallow rooting. Simple water balance models (water in versus necessary water out) have shown the absurdity of this assumption, and subsequent research has shown that these trees can have roots extending up to 17 meters deep into the soil (Nepstad et al. 1994).

One last disclaimer about rainforests is that they are not a single type of forest but a collection of various kinds of forest, including mangrove forests along coasts, swamp forests (*igapo*), mountain forests (montane and **cloud forests**), seasonally flooded forests (*varzea*), peat forests, gallery forests and upland (*terra firme*) forests, and others. What the majority of these forests do have in common is a superabundance of species of flora and fauna.

It is worth noting that there is as yet no single commonly accepted explanation for why so many species can coexist together in the tropics. There are several theories that purport to explain portions of this conundrum in terms of energy flow, ecological **niche** space (ways to make a living), low **disturbance** rates, stable temperatures, and moisture availability, but none is complete. Beyond this, there is even less understanding of how so many species can have come into existence together. One idea that created much interest for a time was the **refugia hypothesis.** The idea was that severe dry periods during the **Pleistocene epoch** caused the vast tropical forests to retreat and break up into isolated refugia, wherein new species formed and then intermixed when the forests recovered. Although appealing, the idea has been largely discredited (Gentry 1989). It has been shown that many of the proposed refugia have never been isolated and that the species distribution patterns, which were the main foundation of the hypothesis, have been shown to be artifacts of sampling (Nelson et al. 1990). Basically, the number of species found is directly related to the amount of effort spent looking for them (Gentry 1989).

Another apparent paradox posed by tropical forests is their ability to support incredibly lush growth of vegetation upon what are usually nutrient-poor and in many cases toxic (aluminum toxicity) soils. Many would-be farmers have discovered that although these forests can support several hundred metric tons of biomass per hectare, they can only support a few years of meager cropping before the topsoil erodes and becomes exhausted of nutrients. It is estimated that only 20% of the soils of the humid tropics can support agriculture using current technology. Furthermore, most of these regions have already been intensively developed (Terborgh 1992). The solution to this apparent riddle is that, in contrast to most forested ecosystems outside of the tropics, the majority of the nutrients are stored in the vegetation, not in the soil. The adaptation that tropical rainforests have made to their existence on soils that have been largely leached of their nutrients by millennia of rains is a superefficient recycling system. In practice, what this means is that any dead biomass, be it animal or vegetable, is rapidly decomposed. Furthermore, the extremely dense packing of fine roots, augmented further still by dense networks of the **hyphae** of symbiotic **mycorrhizal** fungi, rapidly soaks up any nutrients before they can leach into the soils. This recycling can be up to 99% efficient but is more typically 60% to 80% in most tropical forests (Terborgh 1992).

Forest Conversion to Agricultural Lands

The conversion of lush rainforests into productive agricultural lands is not without challenges. Obviously, the first problem is getting rid of the trees. Therefore, the first task is felling roughly 500 good-sized trees per hectare, some of which are immense. If this can be done, the mass of roughly 300 metric tons of vegetation needs to somehow be removed from the desired agricultural plot. Decomposition is fast, but regrowth is even faster, and so the slashed vegetation needs to be removed quickly. Assuming that the land can be adequately cleared, the would-be farmer then has to contend with the poor soils that may only support meager cropping of a limited number of tolerant crops for a few years. Lastly, the agriculturist needs to contend with aggressive and rampant weeds that threaten to quickly overtake the hard-won plot of land. Given this list of challenges, it is not surprising that throughout most of human history, hunter-gatherer groups have predominated.

Eventually, the combination of new technology development and agricultural knowledge provided both the means and the reasons for clearing these forests and planting them with desired plants under a system of shifting agriculture. In shifting agriculture, small areas are cleared and planted for a few years before being allowed to return to forest for extended fallow periods. The one "tool" that stands preeminent in this clearing process, however, is fire. To this day, fire is the main clearance tool because, even in this age of chainsaws and mechanized labor, it is still the fastest and easiest way of converting the large masses of vegetation into nutrient-rich ash. Fire clears and fertilizes the land at the same time. The common terms for this practice are **slash-and-burn** or **swidden agriculture**.

Clearly, slash-and-burn agriculture has been practiced for thousands of years in the tropics, but burning these forests is not a simple matter of lighting a match. Rainforests, as their name implies, are quite wet. In order to burn them, it is therefore necessary to first dry them. The practice of slash-and-burn agriculture commonly proceeds as follows. Within a desired plot, most or all of the trees are felled and then left to dry for several weeks or months. Once cut, the forest canopy no longer shades the ground from the strong tropical sun. Furthermore, the dead vegetation no longer pumps water from the soil to the atmosphere. Thus, temperature and air movement (e.g., winds) increase, both of which act to dry the slashed trees. Seasons are variable across the tropics, but in most regions, there are dry seasons that determine when the practice of slash-and-burn can take place. Therefore, the dry season is also the burning season. Once the slashed materials are judged to be sufficiently dry for burning, the farmer lights the felled vegetation on fire in the late morning or early afternoon so as to get an optimal burn. The farmer's objective is to reduce the pile of debris as much as possible while simultaneously releasing the contained nutrients so that they may act as fertilizer for the soon-to-be-planted crops. If the farmer does not plan well, however, he can either under- or overburn his plot. Effectively this is a function of whether or not he judged the moisture level of the vegetation correctly. If it is too wet, the plot is left as an impenetrable tangle of charred logs, which must usually be abandoned as a total loss. If it was too dry, the material will burn too hot and **volatilize** many of the nutrients, subsequently resulting in poor agricultural productivity and early abandonment.

Newly opened lands are planted with whatever crops are regionally appropriate, and the plot is farmed for as long as it is productive. Lands are frequently reburned every two to three years to reduce weedy invasions (e.g., forest trees). When productivity wanes, the plot may be cycled to pasture for cattle or may be fallowed for a time. If fallowed, the forest will be allowed to regrow and replenish many of the nutrients at the site. Nutrients are accumulated from both continued weathering of soils and atmospheric deposition (while dry or through rainfall). If **fallow cycles** between cuttings are long enough, this is a sustainable practice, but, as population pressures increase, fallow cycles are often shortened, and productivity continues to drop, thus reducing land use options.

Ranching within the tropics is widespread but is variable in size and intensity. As practiced, cattle raising can consist of one or more cows grazing on a few hectares of recently fallowed land or of several thousand head of cattle being moved around to graze various pastures on extensive land holdings. The process for clearing land for pasture is largely the same as for slash-and-burn agriculture, but the clearing often occurs at larger spatial scales as extensive tracts of land are converted to pasture. A newly created pasture may be seeded with grass or simply allowed to regrow. In either case, the pasture will become increasingly overgrown as second-growth vegetation from the forest starts to take over. These pastures can either be cleared by hand with machetes or, as is more commonly practiced, burned again to kill off the forest regrowth.

Non-fire-dependent agriculture using many agroforestry and perennial crops is also an important land use in the tropics. These agricultural practices are potentially more lucrative than ranching or the cultivation of annuals but are at large risk of unintentional burning when incorporated within a matrix of fire-maintained land. Fire-free alternatives to slash-and-burn agriculture and pasture maintenance, such as mechanical mulching, do exist and have been shown to be effective. However, these practices are not yet widespread and may not be practical in many locations.

Deforestation

Deforestation is the name most commonly associated with land cover change in the tropics. It is almost always associated in some way with human land use. Quite simply, it is the removal of forests. The process is not unique to the tropics, having already run its course in much of temperate Asia, Europe, and North America. If, for example, the deforestation that characterized the development of the eastern United States occurred today, as opposed to 100 to 200 years ago, there would be a great outcry at the destruction, which would fragment, degrade, and isolate the forest remnants and cause untold ecological harm.

What makes deforestation in the tropics so stunning is not the extent of the deforestation that has occurred, but the rate at which the forests are disappearing. The combination of burgeoning human populations; the global economy; and the advent of chainsaws, bulldozers, tractors, and other mechanized processes has created deforestation at an unprecedented pace. What would have taken centuries can now be accomplished in a decade or less.

Deforestation in the tropics is widely seen as environmental destruction writ large. In the Brazilian Amazon alone, it is estimated that 15,000 to 20,000 square kilometers of forest are leveled each year (Instituto Nacional de Pesquisas Espaciais 2001). To put that number in some perspective, imagine an area equivalent to the country of Belize, or five times the state of Rhode Island, being cleared each year. Even so, only 15% of the forests of the Brazilian Amazon have been cut so far. Globally, the rate of deforestation is much higher. Although the exact deforestation rate is disputed, it is estimated to be 58,000 to 80,000 square kilometers a year, or roughly three to four times Brazil's acknowledged deforestation rate (Achard et al. 2002; Defries et al. 2002).

One of the myths of tropical deforestation is that, once cut, these forests cannot recover. For a time, it was frequently put forth that once exposed to the drying effects of the sun, the soils would be baked into an impermeable layer of clay. As with many myths, there is some basis in fact for this belief. There are soils in parts of the tropics (**plinthite** or **laterite**) that, if cycled several times between wet and dry, can irreversibly solidify into an **ironstone hardpan** (Van Wambeke 1992). These soils, however, are not the rule for the tropics. Extensive investigation of postclearance forest regeneration has shown that forest recovery is robust and rapid. The caveat to this is that these responses are slowed by increased intensity, duration, and frequency of disturbance. In practice, this implies that the longer an area is kept cleared of forest, the slower it will be to recover. In addition, as clearings become larger, the regeneration process becomes greatly slowed by seed dispersal limitations. Most mature forest tree species in the tropics have large or heavy seeds that cannot disperse very far. Along the margins of tropical forests, forest regeneration may not be possible due to climate restrictions. Forests can maintain themselves under severe seasonal stresses, but they cannot always regenerate under such conditions.

Although agriculture is the land use most associated with deforestation, this change comes in many forms. Throughout much of Southeast Asia, small clearings within forests still predominate in many regions. These subsistence agriculture practices most closely approximate swidden agriculture, with the small number of crops (e.g., manioc, rice) being produced for consumption by the landholder and his or her family.

In more developed areas, however, **production agriculture** has begun to dominate. This can take the form of perennial cash crops or even fully mechanized and fertilized annual cropping. Recently, in Brazil and Bolivia, newly developed soybean plantations have become some of the largest sources of deforestation. New pastures for cattle on large ranches are another major source of deforestation in the tropics.

Plantation forestry, too, is a growing source of deforestation. In West Kalimantan, Indonesia, more than 3.1 million hectares of forest have been allotted to industrial plantations (Curran et al. 2004). Species-diverse forests are replaced with monocultures of highly valuable economic species. These plantations can be for timber, wood pulp, or extracts such as oils and latex. Some of the more important plantation species are oil palm, teak, eucalyptus, and mahogany.

Beyond agriculture, there are a number of other land uses that result in deforestation. Mining operations are one example. Although this activity is locally devastating, its impacts are relatively small in spatial extent compared to agricultural activities. This is not to say that there are not extensive ecosystem ramifications. For example, in the case of gold mining, the main damages come from sediment transport that clogs waterways and from the mercury that is used in gold extraction. Mercury escapes into the environment and can pollute local waterways and the fish that inhabit them. Working up the food chain, this process results in larger and larger concentrations of this poisonous substance in animals that eat the fish, including humans.

Many of the other sources of deforestation can be summed up under the heading of infrastructure development. Roads, utility corridors, hydroelectric dams and their man-made lakes, and general urban expansion all require removal of the native forests. Again, with the exception of dammed rivers, these land uses do not take up a large amount of space *per se*. The land cover changes are effectively permanent, though. Also, the impact of roads, utilities, and towns goes beyond the physical space they occupy. Each of these land uses changes the landscape by influencing what other land use choices are made. Roads provide access to both lands and markets, utilities provide more options for economically viable activities, and towns are the inevitable result of increasing human populations clustering along the roads and utility corridors. Against this backdrop of infrastructure development, it is almost inevitable that agricultural expansion will accelerate, and with it, deforestation rates.

Lost in the discussion of tropical deforestation is the fact that there are more subtle changes occurring to the remaining forests. Forests are degraded through edge effects due to **fragmentation**, are thinned by commercial logging operations, and are highly damaged by escaped wildfires. These activities significantly increase deforestation rates, but, beyond this, they vastly expand the area of forests that have been altered by human activities.

Fragmentation

With each new road and field comes a new forest edge that previously did not exist. Deforestation, whether for agriculture or infrastructure development (e.g., roads, power lines, rail systems, pipelines), results in forest fragmentation. New forest edges are formed, and remnant forests become increasingly affected by disturbance. Globally, Skole and Tucker (1993) estimated that, by 1988, fragmentation and its associated edge effects (e.g., wind exposure, excessive drying, invasive species) had degraded an area of forest 1.5 times the area of that which had been deforested.

Forest fragmentation exposes the remaining forest to increasing levels of disturbance. Within the forest, this can result in biomass collapse, which occurs when the mortality and continued turnover among the trees results in an apparently permanent reduction in the amount of living vegetation (biomass) that the site can support even though it is forested. This effect can extend out to at least 100 to 300 meters from the deforested edge (Laurance et al. 1997). Light penetration through the fractured canopy, and increased levels of woody debris, can make these forests more susceptible to fire. With respect to fire risk, each and every meter of exposed forest is a potential avenue for fire entrance into the forest. Therefore, as a region develops and the forest becomes more fragmented, the risk of forest fires increases.

Two locations in the eastern Amazon illustrate the importance of fragmentation and edge formation for a region's forests. Paragominas is an older frontier that was first settled in the mid- to late 1960s and which is dominated by large ranching and logging interests. Tailândia is a newer frontier area that was created in the 1980s as a settlement project by the Brazilian government (INCRA) for small landholdings. The pattern of fragmentation has differed in the two areas, but the end result has been the same. In both regions, greater than 50% of the remaining forest exists within 300 meters of a forest edge. Fire regularly penetrates the forests of both sites for more than a kilometer, and so virtually all remaining forests in these areas are being periodically burned (Cochrane 2001). These forests now burn at intervals of less than a decade as opposed to periods of centuries or millennia (Sanford et al. 1985; Turcq et al. 1998).

Fire

Fire is an integral part of most agriculture in the tropics. This has been the case since agriculture began in the tropics and likely will continue for the foreseeable future. Until recently, fires in rainforests have been considered either impossible or at best aberrations. Tropical forests are excellent at trapping moisture and keeping the air, and hence potential fuels, very moist even under severe drought stress. Experiments have shown that fires were unable to propagate even after a month without any rainfall (Uhl et al. 1988). It has therefore come as a surprise to many that fire has become such a severe problem in the tropics today. In the 1997–1998 time period alone, fire is conservatively estimated to have burned at least 20 million hectares of tropical vegetation worldwide (Cochrane 2003). The growing pandemic of rainforest fires is now seen as an important consideration in biodiversity conservation, ecosystem function, economic performance, human health, and even the global climate. So what changed?

The 1997–1998 time period included an extreme El Niño event that led to extensive droughts throughout many regions of the tropics. Large fires also occurred in Indonesia during the 1982–1983 El Niño event (Schindele et al. 1989). It is therefore tempting to consider these fires a simple function of anomalous weather. Unfortunately, El Niño weather patterns are not the whole story. While the fires burn spectacularly during some years, eliciting heavy reporting by the media, the fact is that many rainforest fires, even very large ones, are occurring every year.

Now forests that are adjacent to fire-maintained pastures and agricultural lands are at high risk from fire (UNEP 2002). Under the proper climatic conditions, even large tracts of undisturbed forest can burn. Such was the case in Roraima, Brazil, in 1997 and 1998, when fires burned an area of 3.8 to 4.1 million hectares, of which 1.1 to 1.4 million hectares was intact primary forest (Barbosa and Fearnside 1999). Although the El Niño created drought conditions that made these forests highly flammable, the actual fires were caused by the activities of the region's rapidly growing population of rural residents.

The new fire dynamic for many tropical forests is one of frequent fire incursions and increasing fire severity. The first fire in a closed-canopy forest is unimpressive. Except for tree-fall gaps and other areas of unusual fuel structure, the fire spreads as a thin, slowly creeping ribbon of flames only a few tens of centimeters in height. Over much of the burned area, the fire will consume little besides leaf litter. The fire line may move only 100 to 150 meters per day but can keep burning this way for days, weeks, or months. If the weather is cool or a light rain falls, there may be no flames at all. However, large fuel sources, such as fallen logs, can smolder and reignite fires for weeks. Many areas will reburn one or more times as falling leaves from fire-damaged trees repeatedly cover the ground with a flammable litter layer. Logged forest stands are more likely to sustain fires over extended time periods, and to reburn within a single season, than are undisturbed forests that have lower densities of coarse fuels.

The slow advance of tropical fires is what makes them deadly. The size of the flames is not the most important factor affecting tree mortality; it is the amount of time that the flames are in contact with the base of trees that determines the damage they cause. In rainforests, most of the trees have very thin insulating bark layers and are therefore highly susceptible to damage by fire. Bark thickness increases with the diameter of the tree, which explains why smaller trees are more frequently killed by these fires (Uhl and Kauffman 1990).

If a forest reburns within a few years of the initial fire, the fires are much more severe. In recurrent fires, flame lengths, flame depths, spread rates, residence times, and fire line intensities are all substantially higher. When fires recur frequently, large trees have little survival advantage over smaller trees because the changes in fire behavior overwhelm the defenses of even the largest, thickest-barked trees. While the first fire kills mostly small trees, the second fire is just as likely to kill a large tree as a small one (Cochrane et al. 1999).

Fires in selectively logged forests act similarly to those in intact forests except that the first fire may be very intense due to the large amount of slash from logging operations. Therefore, in logged forests that burn, there may be no survival advantage for larger-diameter trees in an initial fire. Therefore, even a single fire in a logged forest will highly degrade the forest and make it much more vulnerable to recurrent fires.

Land cover change is exacerbating the fire problem. Deforestation and conversion to agriculture increases the prevalence and connectivity of flammable ecosystems (e.g., pasture grasses). In the past, agricultural plots and pastures existed as islands of easily flammable vegetation within a sea of largely fire-immune forest. Few forests were exposed to fires, making extensive forest fires rare (UNEP 2002). However, as regions have developed, the forest remnants have become increasingly fragmented and surrounded by large pastures of easily flammable grasses. This fragmentation has exposed more forests to fire-prone edges and has enhanced the overall fire susceptibility of the forests. Connectivity between flammable ecosystems results in a greater chance that each fire will escape into neighboring pastures or agricultural lands and thereby directly increases the economic damages caused by fire while simultaneously exposing more forests to fire. Forest conversion to agriculture affects the regional hydrologic cycle and may increase the tendency for drought and the likelihood of fire (Laurance and Williamson 2001). It is worrisome that forest fires can initiate a positive feedback loop by increasing fire susceptibility, increasing fuel loads, and increasing fire severity. The resultant change in the fire regime can lead to extensive, if unintentional, deforestation (Cochrane et al. 1999).

Logging

In recent decades, the logging industry has boomed in the tropics. The world's appetite for timber and wood products seems to know no limits. Where once only a few high-value species were extracted for commercial markets, now dozens to hundreds of species are being exported. It should be remembered that each hectare of most tropical forest hosts up to hundreds of tree species. By way of contrast, a typical temperate forest in the United States contains only about a dozen species.

In tropical logging operations, forests are rarely clear-cut, since so few of the standing trees have market value. Instead, **selective logging** is practiced, whereby the loggers crisscross the forests, often with bulldozers, looking for valuable trees. Poor felling techniques can kill up to six trees for every one that is cut down, and 40% of all trees can be damaged during logging operations (Uhl et al. 1997). This works out to 20 or more trees being killed or damaged for every one that is extracted. Since so much damage is being done, common logging practices are effectively mining what should be renewable natural resources. In other words, the logs are being cut much faster than they can regrow. Many forests are revisited several times as loggers return to harvest additional tree species that become lucrative when regional timber markets develop. These forests become very degraded. While an intact forest may resist fire encroachment after more than a month without rain, selectively logged forests may become flammable in as few as six to eight rainless days (Uhl and Kauffman 1990).

The vast majority of logging operations in the tropics are wasteful and unsustainable over time. Furthermore, these activities are frequently illegal within the countries where the logging is happening. The shattered remnants of the forests after logging will require many decades, if not centuries, to reachieve their previous stature. This is lamentable because under good management practices—often termed **reduced impact logging** (RIL)—these highly productive forests can support continued timber extraction every 30 years or so (Barreto et al. 1998).

The green gold rush that is the timber boom has already swept over many of the world's tropical forests. The great **dipterocarp forests** of Southeast Asia have largely been logged out. Logging is increasingly moving into what were previously considered forests of marginal economic value, such as those of Papua New Guinea. In Africa, the story is similar. Logging activities are pushing ever deeper into the core of the Congo Basin, bringing with it hunters for the industrial-scale bush meat trade that empties the forests of much of the animal life to provide culinary delicacies for distant city dwellers. Increasingly, logging industries are turning to the last vast stock of tropical timber in South America. Malaysian timber companies have bought rights to huge forest tracts in Suriname, while others have supported road development over the Andes so that the isolated western Amazon forests of Bolivia and Peru can be exploited. The Brazilian Amazon, however, contains more than a third of the world's remaining tropical forests, as well as an extensive river system that makes export of these timber products much easier. Therefore, the Amazon is the biggest prize and is perhaps the last chance for finding sustainable methods of using tropical forests without destroying them.

Development

One person's destruction is another person's development. As governments have striven to turn "unproductive" lands into cash-producing resources, the forests have been leveled. Settlement programs to provide a "land without people to people without land" have measured progress by the number of hectares deforested. Cost and benefit estimations of such activities have varied widely depending on whose economic analyses were used. In many cases, international aid or development groups have financed these projects, only later to realize the extensive environmental costs involved. Forests have given way to grasslands, and biomass accumulation is no longer measured in metric tons of vegetation but in kilos of beef. What one group sees as chaotic destruction, another group sees as the foundation for economic expansion and social pacification. Slashed forests equal land for settlement. Destruction and displacement of indigenous species, including humans, is the price for buying a chance at a future for millions of rural poor, and for rich speculators too. In short, the tropics are the latest, and perhaps last, frontier, where a more powerful people is dominating previous inhabitants and destroying vast ecological riches in the pursuit of wealth, without realizing that the natural resources are not limitless. It is a story that has been told time and again throughout human history.

The history of the Brazilian Amazon illustrates this well. In the 1960s and 1970s, the military government sought to secure its hold on the vast Amazon region through development and settlement to fend off any potential land claims from other nations. By cutting the Transamazon highway through the midst of the tractless jungle, Brazil demonstrated its dominance. It also provided a convenient outlet for the resettlement of millions of people from the burgeoning poor populations of the rest of Brazil. After an initial planned approach, however, the subsequent 30 years of settlement expansion and development of the Amazon have been allowed to proceed in a haphazard, uncontrolled fashion (Laurance et al. 2001).

Now, though, the Brazilian government is again trying to seize control of Amazonian development through an ambitious $40 billion infrastructure expansion program called "Advance Brazil." Under this program, new power generation and distribution will complement river channelization, road improvement, and development to greatly expand the economic potential of the region. For example, soybean production from the southern Amazon will now be able to be shipped directly to the world's markets via river ports instead of by long, slow, and expensive trucking to the south of Brazil. Timber, too, will be much easier to process and ship. The infrastructure will also greatly expand the area of forest accessible for colonization, as well as the economic opportunities for the 20 million residents of the Brazilian Amazon. Not too surprisingly, this has led to dire predictions that, unless development patterns change, deforestation rates in the Brazilian Amazon could jump to 25,000 square kilometers per year, and as little as 5% of the forest may remain in pristine condition 20 years from now (Laurance et al. 2001).

Conservation and Sustainability

Globally, there has been growing concern for the long-term conservation of rainforests and the diversity of life that they contain. While large regions of the Amazon forest remain intact, the remnants of most tropical forests are under immediate threat. In some countries—Haiti, for example—the forests have been virtually eliminated. In other regions, such as west Africa, the last few forests are being rapidly cleared. The largest rates of deforestation are occurring in Southeast Asia (Achard et al. 2002). The forests of Indonesia are of particular concern due to the rampant and out-of-control logging operations (Smith et al. 2003) that are stripping the country of its natural wealth and leaving tremendous fire hazards in their wake. Even in countries such as India, where a fairly strong conservation system of parks and reserves exists, elevated disturbance levels are slowly eroding biodiversity, if not the total forest cover.

Tropical countries have legally established many "**paper parks**" for conservation purposes, but there are rarely the monetary and human resources necessary to manage, monitor, and protect them. To date, despite such weaknesses, these parks have been remarkably effective at minimizing deforestation, albeit less so regarding logging (Bruner et al. 2001). However, without any long-term strategy for meeting the needs of regional populations, it is only a matter of time before decreasing stocks of available forests outside the parks force development pressures to encroach upon them. In short, if the conservation parks are to be sustainable refuges for tropical biodiversity, then the areas outside the parks must be able to sustainably support their human populations.

Brazilian conservation plans illustrate one attempt at achieving this parity. With timber stocks in Southeast Asia and Africa dwindling, demand for Amazonian timber production continues to increase. Plantation forests cannot supply the world's timber needs for the foreseeable future. With its vast timber stocks and expanding infrastructure, the Amazon is therefore well positioned to dominate the tropical timber trade in the 21st century. Already, the breadth of area being exploited for timber is growing dramatically. In 1996, only 500,000 hectares of standing forest were being logged each year, but by 1999, this rate had grown to nearly 1.2 million hectares per year (Matricardi 2003). Almost all of this logging has occurred on either private or unclaimed government lands. Of the logs arriving at sawmills, 50% or more have been illegally harvested, and the great majority (95%) come from unmanaged logging operations that generate severe impacts on forest structure, put excessive pressure on high-value species, and render the forests susceptible to fires (Veríssimo and Cochrane 2003).

Until recently, there have been few alternatives to predatory logging other than a few forest management experiments. Interest in stable and sustainable forestry has grown, though. There are now in excess of 1 million hectares of forest that have been certified by the Forest Stewardship Council (FSC) as providing "green wood products." The move to sustainable forest use has been accelerated by Brazil's new commitment to vastly expanding its national forest system, **Flonas.** By law, Flonas must be managed sustainably. Before this current policy initiative, Flonas comprised less than 2% of the Amazon Basin. Even if all of these forests were used for sustainable production, they could provide no more than 11% of the demand for Amazonian timber. In fact, approximately 70 million hectares (14%) of Brazil's Amazonian forests need to be brought into well-managed production in order to satisfy the demand for wood. The new forest policy being implemented by the Brazilian government, based on well-managed production, is now expanding the national forest system to 50 million hectares in the Amazon (Veríssimo et al. 2002).

Establishing sustainable production forests to provide for long-term tropical timber extraction is only one step toward achieving a truly sustainable management system and only one part of a larger conservation strategy. It is, however, the first step that needs to be taken. It is critical that these forests be established now before economic and social factors make such action politically infeasible. Research shows that there are still substantial amounts of forest that can be incorporated into Flonas with little or no conflict with current protected areas or human populations.

The current perception of timber superabundance is generating transient (boom-and-bust) economic activity as logging operations move with the frontier. Strategic expansion of the Flonas system can contribute to both biodiversity conservation and economic stability in the Amazon by constraining unsustainable development activities. The establishment of a more expansive system of Flonas could also be instrumental in reducing the negative impacts of such development programs as Advance Brazil and could dramatically alter the dire predictions of future forest destruction and degradation based on current predatory models of development (Laurance et al. 2001).

Combining **biodiversity conservation** and the best forest management practices is necessary to establish truly sustainable production and achieve conservation goals. Protection of areas of high biological significance will require the creation of a mosaic of conservation areas that combines Flonas (sustainable use) with parks and biological reserves (full protection). The Brazilian government has recently announced plans to protect biodiversity by turning 10% of the Amazon into fully protected parks and biological reserves. In this system, Flonas would form a buffer zone around parks and reserves. In addition, Flonas could also provide corridors for movement of species between core protection areas. These new initiatives, combined with existing protected lands (indigenous lands, extractive reserves, etc.), will increase the amount of forest protected from deforestation to roughly 42% of the Brazilian Amazon. If these programs are fully implemented, Brazil will become the world's foremost nation in conserving natural resources and biodiversity. These efforts mark the best hope yet for achieving sustainable forestry, biodiversity conservation, and human well-being in the tropics.

REFERENCES

Achard, F., H. D. Eva, H. Stibig, P. Mayaux, J. Gallego, T. Richards, and J. Malingreau. 2002. Determination of deforestation rates of the world's humid tropical forests. *Science* 297:999–1002.

Barbosa, R. I., and P. M. Fearnside. 1999. Incendios na Amazonia Brasileira: Estimativa da emissão de gases do efeito estufa pela queima de diferentes ecossistemas de Roraima na passagem do evento Ël Niño (1997/98). *Acta Amazonica* 29:513–34.

Barreto, P., P. Amaral, E. Vidal, and C. Uhl. 1998. Costs and benefits of forest management for timber production in eastern Amazonia. *Forest Ecology and Management* 108:9–26.

Bruner, A. G., R. E. Gullison, R. E. Rice, and G. A. B. da Fonseca. 2001. Effectiveness of parks in protecting tropical biodiversity. *Science* 291:125–28.

Cochrane, M. A. 2001. Synergistic interactions between habitat fragmentation and fire in evergreen tropical forests. *Conservation Biology* 15:1515–21.

———. 2003. Fire science for rainforests. *Nature* 421:913–19.

Cochrane, M. A., A. Alencar, M. D. Schulze, C. M. Souza Jr., D. C. Nepstad, P. Lefebvre, and E. Davidson. 1999. Positive feedbacks in the fire dynamic of closed canopy tropical forests. *Science* 284:1832–35.

Curran, L. M., S. N. Trigg, A. K. McDonald, D. Astiani, Y. M. Hardiono, P. Siregar, I. Caniago, and E. Kasischke. 2004. Lowland forest loss in protected areas of Indonesian Borneo. *Science* 303:1000–1003.

DeFries, R. S., R. A. Houghton, M. C. Hansen, C. B. Field, D. Skole, and J. Townshend. 2002. Carbon emissions from tropical deforestation and regrowth based on satellite observations for the 1980s and 1990s. *Proceedings of the National Academy of Sciences* 99:14256–61.

Gentry, A. H. 1989. Speciation in tropical forests. In *Tropical forests: Botanical dynamics, speciation and diversity.* Edited by L. B. Holm-Nielsen, I. C. Nielsen, and H. Balslev. San Diego: Academic Press Limited.

Instituto Nacional de Pesquisas Espaciais. 2001. *Monitoramento da Floresta Amazônica Brasileira por Satélite: 1998–2000.* São Paulo, Brazil: São José dos Campos. http://www.inpe.br/Informacoes_Eventos/amazonia.htm.

Laurance, W. F., M. A. Cochrane, S. Bergen, P. M. Fearnside, P. Delamônica, C. Barber, S. d'Angelo, and T. Fernandes. 2001. The future of the Brazilian Amazon: Development trends and deforestation. *Science* 291:438–39.

Laurance, W. F., S. G. Laurance, L. V. Ferreira, J. Rankin-de Merona, C. Gascon, and T. E. Lovejoy. 1997. Biomass collapse in Amazonian forest fragments. *Science* 278:1117–18.

Laurance, W. F., and B. Williamson. 2001. Positive feedbacks among forest fragmentation, drought, and climate change in the Amazon. *Conservation Biology* 15:1529–35.

MacDonald, G. M. 2003. *Biogeography, space, time and life.* New York: John Wiley & Sons.

Matricardi, E. A. T. 2003. Multi-temporal assessment of selective logging using remotely sensed data in the Brazilian Amazon. M.A. Thesis. Michigan State University.

Nelson, B. W., C. A. C. Ferreira, M. F. da Silva, and M. L. Kawaski. 1990. Endemism centres, refugia and botanical collection density in Brazilian Amazonia. *Nature* 354:714–16.

Nepstad, D. C., C. R. de Carvalho, E. Davidson, P. Jipp, P. Lefebvre, G. H. Negreiros, E. D. da Silva, T. Stone, S. Trumbore, and S. Vieira. 1994. The role of deep roots in water and carbon cycles of Amazonian forests and pastures. *Nature* 372:666–69.

Sanford, R. L., J. Saldarriaga, K. Clark, C. Uhl, and R. Herrera. 1985. Amazon rainforest fires. *Science* 227:53–55.

Schindele, W., W. Thoma, and K. Panzer. 1989. The forest fire in East Kalimantan. Part I: The fire, the effects, the damage and technical solutions. FR-Report No. 5.

Skole, D., and C. J. Tucker. 1993. Tropical deforestation and habitat fragmentation in the Amazon: Satellite data from 1978 to 1988. *Science* 260:1905–10.

Smith, J., K. Obidzinski, Subarudi, and I. Suramenggala. 2003. Illegal logging, collusive corruption and fragmented governments in Kalimantan, Indonesia. *International Forestry Review* 5(3):293–302.

Terborgh, J. 1992. *Diversity and the tropical rain forest.* New York: Scientific American Library.

Turcq, B., A. Sifeddine, L. Martin, M. L. Absy, F. Soubies, K. Suguio, and C. Volkmer-Ribeiro. 1998. Amazonia rainforest fires: A lucustrine record of 7000 years. *Ambio* 27:139–42.

Uhl C., P. Barreto, A. Veríssimo, E. Vidal, P. Amaral, A. C. Barros, C. Souza, J. Johns, and J. Gerwing. 1997. Natural resource management in the Brazilian Amazon. *BioScience* 47:160–68.

Uhl, C., and J. B. Kauffman. 1990. Deforestation, fire susceptibility, and potential tree responses to fire in the eastern Amazon. *Ecology* 71:437–49.

Uhl, C., J. B. Kauffman, and D. L. Cummings. 1988. Fire in the Venezuelan Amazon 2: Environmental conditions necessary for forest fires in the evergreen rainforest of Venezuela. *Oikos* 53:176–84.

UNEP and M. A. Cochrane. 2002. *Spreading like wildfire—tropical forest fires in Latin America and the Caribbean: Prevention, assessment and early warning.* Mexico City, Mexico: United Nations Environment Programme. www.rolac.unep.mx/dewalac/eng/fire_ingles.pdf.

Van Wambeke, A. 1992. *Soils of the tropics.* New York: McGraw-Hill.

Veríssimo, A., and M. A. Cochrane. 2003. Management and conservation of tropical forests: Brazil's bold initiative in the Amazon. *ITTO Tropical Forest Update* 13(3):4–6.

Veríssimo, A., M. A. Cochrane, and C. Souza Jr. 2002. National forests in the Amazon. *Science* 297:1478.

Wallace, R. A. 1876. *The geographical distribution of animals. With a study of the relations of living and extinct faunas as elucidating the past changes of the earth's surface.* New York: Harper & Brothers.

SUGGESTING READINGS

Cochrane, M. A. 2003. Fire science for rainforests. *Nature* 421:913–19.

Laurance, W. F., M. A. Cochrane, S. Bergen, P. M. Fearnside, P. Delamônica, C. Barber, S. d'Angelo, and T. Fernandes. 2001. The future of the Brazilian Amazon: Development trends and deforestation. *Science* 291:438–39.

Laurance, W. F., and B. Williamson. 2001. Positive feedbacks among forest fragmentation, drought, and climate change in the Amazon. *Conservation Biology* 15:1529–35.

Stokstad, E. 2003. "Pristine" forest teemed with people. *Science* 301:1645–46.

Veríssimo, A., and M. A. Cochrane. 2003. Management and conservation of tropical forests: Brazil's bold initiative in the Amazon. *ITTO Tropical Forest Update* 13(3):4–6.

HAPTER

3

The Sweet Earth

A
)GEOCHEMISTRY
ERSPECTIVE OF
OPICAL FOREST
SOILS

Deborah A. McGrath
and
C. Ken Smith

Introduction

We might think of soil as simply lifeless "dirt," but there are major differences between soil and the stuff we sweep up off the floor. Soil is a dynamic natural body composed of **mineral** and **organic matter** that provides a medium for plant growth (Brady and Weil 2002). Equally important, soil furnishes habitat for multitudes of living organisms. On average, a gram of forest soil contains billions of individual microorganisms representing some tens of thousands of species. This makes a simple handful of soil one of the most biologically diverse **ecosystems** on Earth (Wilson 2002). Soils also cleanse our water supply, assimilate organic wastes that are recycled by microorganisms, and provide a terrestrial sink for elements, such as carbon, that might otherwise pollute our atmosphere (Brady and Weil 2002). Soils physically support our homes and cities and produce an ever-increasing amount of food and fiber. In short, soils contribute critically to the economic wealth of human societies and to the healthy functioning of our biosphere (Richter and Markewitz 2001).

While humans have derived great benefit from soils over the millennia, we rarely stop to think about how our activities degrade a soil's capacity to perform the functions upon which we are utterly dependent. Land use change—in particular, forest clearing for agricultural crops, pasture, and housing developments—drastically alters the natural inputs of matter, water, and energy that sustain soil fertility. The problem is of particular concern in the tropics, where forest clearing and conversion has accelerated at unprecedented rates over the past two decades. The soils underlying tropical forests are often (but not always) nutrient poor, rendering the entire ecosystem dependent upon a tight recycling of elements among the atmosphere, biosphere, and soil, a process referred to as **nutrient cycling** (see figure 3.1). In essence, forest nutrient cycles can be viewed as a "subsidy from nature" in that they maintain ecosystem productivity for free. Some of the soil maintenance functions provided by a forest can be replaced, at a cost, by intensive management, which might include the use of **amendments** such as synthetic fertilizers or cover crops. However, such practices may not be within the means of resource-limited farmers, especially those living in developing countries where some types of soil amendments are difficult to obtain.

Vegetation physically protects the soil from erosion so that topsoil is not transported off the site by wind, water, or gravity, and into streams and lakes. Soil erosion is particularly destructive because the lost soil cannot be replaced rapidly by soil formation, which takes place over millions of years (Brady and Weil 2002). Throughout human history, thriving ancient civilizations have been brought down by erosion-induced soil degradation (Perlin 1989). From this perspective, soil is a nonrenewable resource (Lal 1990). Thus, in the absence of careful management, forest clearing and subsequent disruptions in elemental cycles can lead to the degradation of

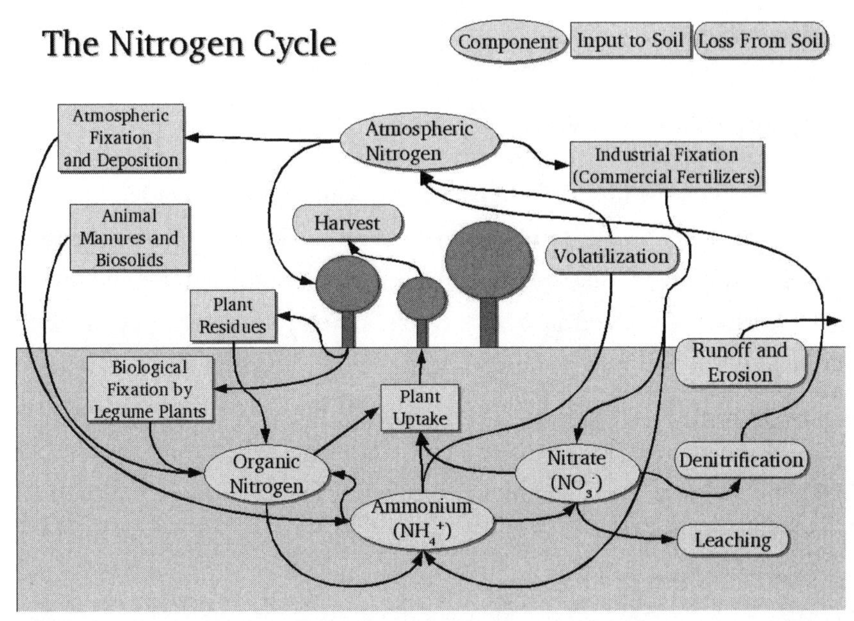

Figure 3.1. A typical forest nitrogen cycle demonstrating important nutrient cycling processes that maintain ecosystem productivity. Adapted from a figure presented by Mississippi State University (MSU), www.msucares.com/crops/soils/.

soil and water quality, a decline in agricultural productivity and the capacity of soils to store atmospheric pollutants, and the loss of one of our most basic resources.

In this chapter, we begin our discussion about the impact of forest clearing on tropical soils by first outlining general concepts about soil and its formation and about nutrient-cycling processes that maintain soil productivity. We then examine how land use change in the tropics affects important chemical, physical, and biological properties that determine soil quality and, in turn, greatly affect a soil's capacity to support human and natural ecosystems. We also attempt to dispel some common misconceptions by discussing some "myths" about tropical forest soils. We end the chapter by exploring some recent trends in soil science research, especially in the field of **biogeochemistry**, and discussing how we can apply what we have learned from these studies to improving our management of soils in a world of rapid land use change.

Scientific Concepts

SOIL AND HOW IT IS FORMED

To appreciate how rapidly land use change can alter or degrade soil quality, we must first understand the processes that contribute to soil formation and the interacting factors that influence its properties. Fundamentally, soils are derived from rocks, which are composed of minerals. The **lithosphere** is continually altered both physically and chemically as it is exposed to the atmosphere, hydrosphere, and biosphere—a process known as **weathering.** Rocks are broken down physically into progressively smaller particles, and the minerals constituting rocks are transformed chemically as they react with air, water, organisms, and other minerals (Bridges 1997).

Weathering produces soils that vary in **texture** and chemical composition. Soils with a sandy texture have greater **pore space** than clayey soils made up of finer particles. Soil porosity is an important component of the soil environment because it determines how easily oxygen and water can move through the soil as they are exchanged among living organisms and the atmosphere. An ideal soil would have approximately 50% pore space, holding about equal amounts of air and water. The remaining half would be composed of roughly 45% mineral solids and 5% organic matter. Highly compacted soils, with **bulk densities** ranging from 1.5 to 2.0 grams of soil per cubic centimeter, have restricted pore space and resist root penetration more than those with lower bulk densities (Brady and Weil 2002).

Soil color provides an indication of the minerals in a soil and the weathering processes affecting it. In a soil profile, which can be seen by looking at a road cut, the horizontal layers delineated by differences in color and texture are called **horizons** (see figure 3.2). The layer of accumulated organic matter (mostly fallen plant parts) on top of the mineral soil is referred to as the **O horizon**. As this horizon undergoes **decomposition**, it forms **humus**. The O horizon accumulates at greater depths in soils with slower rates of organic matter decomposition, such as those found in cold

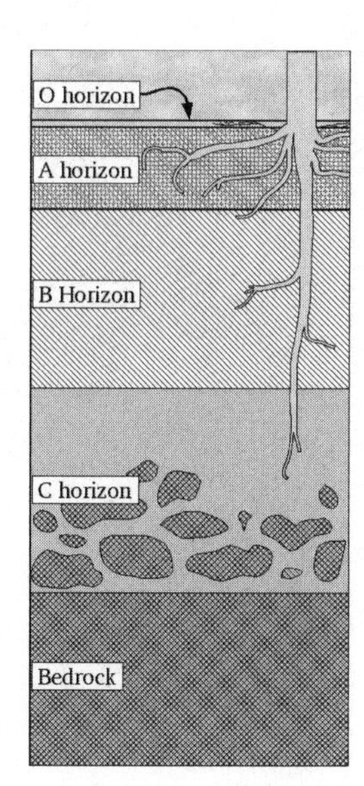

Figure 3.2. A soil profile demonstrates horizons differentiated by the chemical and physical characteristics of the soil at a particular depth. Note that root growth is concentrated in the more mineral-rich A horizon. In tropical soils, rootlets often grow into the organic (O) horizon above the mineral soil surface.

or dry climates. The **A horizon**, often the most "nutrient-rich" layer of soil, is characterized by a darker color resulting from the mixing of humus from the O horizon. In the **B horizon**, fine-textured clays washed down from the A horizon accumulate, giving this layer a more distinctive structure, or aggregation of particles. The **C horizon** refers to the soil layer perched just above the **bedrock**.

Two of the most weathered **soil orders** (see table 3.1) commonly found throughout the tropics are deep red, yellow, or orange in color because they contain high concentrations of metals such as aluminum and iron. The profile of these soils is very deep and often high in clay as a result of the extensive weathering they have undergone. Frequently the O horizon is thin due to the rapid rates of organic matter decomposition that take place in hot, humid environments.

Soil pH indicates the degree of acidity or alkalinity of a soil and significantly affects the environment of soil organisms. Soil pH also affects nutrient availability to plants by reducing **cation exchange capacity** (CEC). Base metal cations, such as Ca^{2+}, Mg^{2+}, and K^{+}, are important plant nutrients that are attracted to negatively charged surfaces of soil particles and

Table 3.1. A simplified summary of the world's twelve soil orders, based upon the USDA soil taxonomy system that groups soils based upon properties that reflect their development. It should be noted that soils are characterized in different countries by different taxonomic systems, and classified based on differing criteria, such as developmental state, climatic region, and management purposes. This table is meant to provide a general idea of the different soil orders and their approximate occurrence throughout the tropics.
Adapted from Brady and Weil (2002), Bridges (1997), Van Wambeke (1992) and Sanchez (1976).

SHARED CHARACTERISTICS	SOIL ORDERS INCLUDED	SPECIFIC PROPERTIES OF EACH ORDER	APPROXIMATE % OF TROPICAL SOILS*
Highly weathered	Alfisols	Soils with an argillic (clay-enriched) horizon; moderate to high base cations	55
	Oxisols	Deep with an oxic horizon containing iron oxides, few minerals, very low fertility	
	Ultisols	Soils with an argillic horizon and low base saturation, low fertility	
Light-colored base rich soils	Aridisols	Soils of deserts/semi-desert areas (dry conditions $>$ 50% of year), no dark horizon	19
Shallow soils or alluvial (water deposited) soils	Entisols	Young soils with a weakly developed profile	18
	Inceptisols	Moderately developed soils	

SHARED CHARACTERISTICS	SOIL ORDERS INCLUDED	SPECIFIC PROPERTIES OF EACH ORDER	APPROXIMATE % OF TROPICAL SOILS*
Dark-colored base rich soils	Mollisols	Soils with a dark A horizon and high base cation content	6
	Vertisols	Soils with >30% clay in all horizons, deep cracks (50 cm), shrink-swell properties	
Moderately weathered and leached	Andisols	Soils derived from volcanic materials with a high cation exchange capacity	1
	Spodosols	Sandy soils with spodic horizon that contains Fe and/or Al oxides and humus accumulation	
Organic soils	Histosols	Soils with >30% organic matter	1
Frozen soils	Gelisols	Soils of cold regions with permafrost	0

*Based upon Van Wambeke (1992) and Sanchez (1976)

organic matter. Frequently, acidic soils have a low CEC because many of the negatively charged sites on particles are occupied by H^+, leaving base cations free in the soil solution and susceptible to loss by downward movement of water. In highly acidic soils, with pH values ≤ 4.5, trivalent aluminum (Al^{3+}) becomes increasingly more soluble, often leading to aluminum toxicity in plants. Root growth, in particular, can be severely restricted by the presence of Al^{3+} in the soil solution, which in turn reduces plant absorption of water and nutrients. Another common problem in acidic soils is that phosphate ($H_2PO_4^-$), an important plant nutrient, reacts with aluminum and iron (Fe^{3+}) minerals, forming a complex that holds it very tightly onto the soil and makes it unavailable for plant uptake.

FACTORS INFLUENCING SOIL FORMATION

Parent material, or bedrock, refers to the type of substrate from which a soil is weathered and greatly influences both its texture and its initial mineral composition. Soils can be derived in place from underlying rock or can be formed from sediments transported and deposited by agents such as gravity (e.g., down a mountain), rivers, glaciers, and wind. For example, soils derived from quartz-rich materials, such as sandstone, may be coarse grained and less abundant in a variety of plant nutrients because quartz is nothing but silica and oxygen and thus resists physical breakdown. In contrast, limestone often contains clays and sediments deposited by marine organisms over millions of years and thus produces a fine-textured, more mineral-rich soil.

Climate, principally precipitation and temperature, determines the nature and intensity of soil weathering that occurs over a large geographic region. Freeze-thaw cycles typical of colder climates promote the physical breakdown of parent material as water and ice expand and contract, creating small fissures in rocks. The dissolution of rock minerals is more rapid in regions with high rainfall and warm climates. Carbon dioxide (CO_2) dissolved in rainwater forms carbonic acid, which is a weak acid but an important weathering agent. The mass movement of precipitation through soil leaches nutrients down the soil profile, often as far as the water table. This elemental loss acidifies soils and further contributes to their chemical weathering. In addition, metabolic activity in bacterial and fungal decomposers is stimulated by heat and humidity, resulting in higher rates of organic matter decomposition in the tropics.

Time influences soil formation by determining the extent to which a soil has been weathered. Some soils are derived in place from ancient rock formations that have undergone such extensive weathering that the soil is depleted of mineral nutrients needed by plants and other organisms.

Many soils of the Amazon Basin, for example, were formed from **Precambrian** rocks uplifted and initially exposed to weathering processes well over 550 million years ago (Sombroek 1984). These soils are deep, clayey, acidic, and have a low CEC due to the extensive periods over which weathering processes have altered the soil. In contrast, some soils on islands in the Pacific are derived from the mineral-rich ash and lava of recently erupted volcanoes, and they are typically richer in most plant nutrients (Bridges 1997; Van Wambeke 1992).

Topography, or the configuration of the land surface, influences the stability and thickness of a soil. Steep slopes encourage rapid soil loss by erosion, inhibit deep penetration by rainfall, and produce the thin drought-prone soils typical of many mountainous regions. Smaller soil particles, along with organic matter, accumulate in concave depressions below, creating deeper, more fine-grained and nutrient-rich soils, which is why farming communities are often located in valleys, not on the ridges above.

Living organisms, or biota, critically influence soil formation in a number of ways. In addition to holding soil in place, vegetation stores nutrients. The minerals taken up from soil by roots are assimilated into the plant body as biomass, thus reducing the mineral loss that occurs by leaching and erosion. Important plant nutrients include calcium (Ca^{2+}), magnesium (Mg^{2+}), potassium (K^{+}), nitrogen in the forms of ammonium (NH_4^{+}) and nitrate (NO_3^{-}), phosphorus as phosphate ($H_2PO_4^{-}$), and sulfur as sulfate (SO_4^{-}). Tree trunks store a lot of calcium, while considerable amounts of nitrogen are contained in leaves. Plant roots contribute to the chemical weathering of soil by exuding organic acids that dissolve minerals and feed microorganisms. As primary producers, plants add organic matter to the soil in the form of "litter," or fallen dead parts such as leaves, twigs, and fruits. Organic matter increases the capacity of soil to retain moisture and cations, and provides food for soil fauna.

Living organisms also influence soil through **symbiosis**. In exchange for **photosynthate** produced by the host plant, **mycorrhizal** fungi secrete substances that protect the plant from disease and increase its absorption of nutrients from the soil. Bacteria living in root nodules of leguminous plants perform **nitrogen fixation**, converting atmospheric nitrogen (N_2) into ammonium, a form that can be assimilated by their host. The bacteria receive carbohydrates produced during **photosynthesis**, and host plants gain access to more nitrogen. Indirectly, this element is added to the soil with the shedding and decomposition of nitrogen-enriched plant tissues. This is an important process because the industrial conversion of N_2 to a form that can be taken up by plants requires an enormous input of energy from fossil fuels, making synthetic fertilizers costly.

NUTRIENT CYCLING MAINTAINS SOIL PRODUCTIVITY

Before nutrients in **detritus** are again available for plant uptake, decomposing litter must undergo **mineralization** to release ions back into the soil solution (see figure 3.1). As microorganisms break down carbohydrates in organic matter to fuel metabolic processes, organic carbon (C) is oxidized and released as carbon dioxide (CO_2), while the other elements (N, P, K, Ca, Mg, S) are either assimilated by the organisms or released as ions available for plant uptake.

Soil **food webs** (see figure 3.3) demonstrate the important interdependent activities of soil flora and fauna that maintain nutrient cycling and soil health. Primary consumers, such as bacteria, fungi, **nematodes**, and **amoebae**, feed upon decaying fine roots and other detritus. Earthworms actually consume mineral soil, grinding it down and releasing it as nitrogen-rich castes. Termites break down **cellulose** and **lignin** in woody debris, releasing the minerals tied up in wood back into the soil. Termites also redistribute clay, organic matter, and minerals when they build their nests, strongly influencing the texture and chemical composition of soils, especially those in tropical dry woodlands (Menaut et al. 1995). These and other insects provide food for secondary consumers, such as birds or anteaters. Larger soil fauna, such as moles that prey on earthworms, burrow and tunnel, further mixing and aerating the soil. Soil food webs are a critical component of the soil environment, and thus a healthy soil is one that supports a diversity of

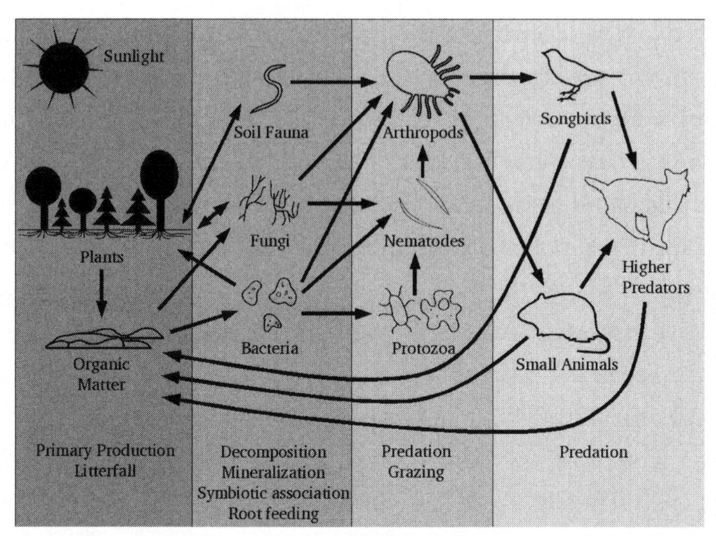

Figure 3.3. Soil food webs are critical to maintaining soil productivity. Adapted from a figure presented by the Bureau of Land Management (BLM), www.blm.gov/nstc/soil/foodweb/.

living organisms, who in turn maintain soil productivity by recycling nutrients back into the soil.

Not surprisingly, rates of organic matter decomposition and mineralization depend upon soil factors that control populations of microorganisms, such as aeration, moisture, temperature, acidity, and the nutrient content of the organic matter itself. Vegetation removal leaves soil exposed to drying elements such as the sun and wind, thus drastically altering the microenvironment of soil fauna. In seasonally dry tropical forests, decomposition and mineralization rates decline during the dry season as microbial populations die or become dormant (Cuevas 1995). In the lowland humid tropics, however, a rise in soil temperatures following clearing may increase the activity of microbial populations. Soil disturbance, such as cultivation, increases decomposition by promoting aeration and exposing new surfaces to microbial attack.

Mineralization rates also depend on the type of plant material being broken down. For example, woody debris contains very little nitrogen compared to leaves, which is one of the reasons why leaf litter breaks down so much more rapidly than twigs and branches. In general, pine needles decompose more slowly than leaves from hardwood trees growing in the same soil because pine litter has more carbon-containing defense compounds that better resist decay. Thus changes in a plant community occupying a site may produce distinct differences in nutrient-cycling dynamics that can affect soil productivity.

Another important concept in nutrient cycling is a nutrient budget, which quantifies the content of a specific element (for example, nitrogen or phosphorus) throughout an ecosystem (soil and plant biomass), as well as how much of the nutrient is added or taken away from the system. Forests are essentially closed systems in which minerals from the soil are taken up by plant roots, incorporated into plant tissues, and returned to the soil as litter. Some nutrients may be added to the system with rainfall, which is often enriched with nitrogen or sulfur from dust, smog, and smoke. Nutrient losses in forests occur with herbivory, when animals consume plant material and then leave the forest. However, these inputs and outputs are relatively minor compared to those that occur in agricultural systems, tree plantations, or harvested forests, when significant amounts of biomass and nutrients are removed with food crops or timber. As you might imagine, productivity in any managed system, especially those growing in highly weathered and nutrient-depleted soils, cannot be sustained over long periods unless outputs are balanced by large inputs, such as fertilizers.

The Effects of Land Use Change on Soil

Soils are not uniformly affected by **deforestation**. The impact of forest clearing on soil properties depends upon how the forest is cleared, the type of land use that replaces it, and the management and duration of that land use. In this section, we discuss two widely practiced land uses that drive forest conversion throughout the tropics and have a profound effect on soils: slash-and-burn agriculture and pasture creation for livestock. We focus our discussion on the conditions under which the effects of these practices are most severe, such as in soils that are highly weathered initially, or on land that is particularly vulnerable to erosion.

SLASH-AND-BURN AGRICULTURE

An image of tropical deforestation frequently portrayed by the media is the decimation of rainforests by poor farmers in developing countries who cut and burn native vegetation to clear lands for agriculture, a practice referred to as **slash-and-burn agriculture.** While the image of charred rainforest fragments surrounding a newly cleared agricultural field may be shocking, there are very rational reasons for practicing this type of agriculture. When forest biomass is cut down and burned, the organic carbon is oxidized (releasing CO_2 as a by-product), and the nutrients in the plant biomass are mineralized and deposited on top of the soil as ash. Thus, burning not only removes unwanted vegetation from the land but also provides a pulse of mineral nutrients that fertilizes crops. The addition of base-forming cations (Ca^{2+}, Mg^{2+}, and K^{+}) also raises the pH of the soil and buffers it against future changes in pH. Burning may also be used to prepare the field the following year to eliminate pests such as weeds. As such, slash-and-burn agriculture turns an acid forest soil into a productive agricultural field for two to three years (Ewel et al. 1981; Juo and Manu 1996).

Once a site is cleared of vegetation, it is at risk for erosion, especially as fields are usually prepared for cropping at the onset of the rainy season. Worldwide, it is estimated that up to 2 million hectares of arable land are lost annually to severe soil erosion (Lal 1990). By washing away topsoil, erosion depletes soil of nutrients, decreases the rooting depth of plants, and reduces soil water storage. The eroded soil is often deposited as sediments in waterways, such as streams or rivers, degrading water quality and endangering aquatic animal populations. **Sedimentation** also decreases the capacity of reservoirs and hydroelectric facilities and blocks waterways (Lal 1990).

The severity of erosion depends upon climate, land use, topography, and soil properties. Newly cleared areas in the tropics are especially susceptible to erosion because of the quantity and intensity of rainfall received in these regions. Cultivation in mountainous regions, such as those found in the Himalayas, Ethiopia, Haiti, and Costa Rica, often produces severe erosion as rainfall rapidly washes soil down denuded slopes. In arid regions, such as Sahelian Africa, where vegetation is sparse, wind erosion results in the loss of both mineral-rich ash and topsoil following a slash-and-burn (Maass 1995). Soil erosion following forest clearing degrades even the most fertile tropical soils in regions such as India and Eastern Africa (Lal 1990).

Burning destroys the soil O horizon, thus greatly reducing the many important services (faunal habitat and food, soil protection, water retention, and nutrient exchange) it renders to an ecosystem. In addition to organic matter loss, much of the nitrogen in plant tissues is destroyed during a burn through the process of volatilization, which transforms organic nitrogen to a gas (NO_x), returning this element to the atmosphere. Studies in tropical forests have shown that burning can result in a loss of 50% to 90% of the nitrogen once contained in the combusted plant material (Kauffman et al. 1998).

Following clearing, decomposition of organic matter proceeds more rapidly with a rise in surface soil temperatures, especially on burned sites. As decomposition rates accelerate in response to nutrient inputs from ash and aeration from cultivation, soil organic matter diminishes rapidly. Often, rates of decomposition and mineralization outpace the capacity of young crop plants to take up nutrients from the soil, leaving ions susceptible to loss as they are leached beyond the rooting zone. Studies have shown that increased nitrogen mineralization rates following a burn make nitrate-N very susceptible to loss (Matson et al. 1987). Interestingly, some studies demonstrate a decline in nitrogen mineralization several years after conversion, perhaps due to reduced nitrogen inputs to the converted land use as compared to that cycled in forests (Neill et al. 1999).

The longevity of soil nutrient pulses following clearing depends upon soil texture, remaining soil organic matter, the extent to which soil CEC is pH dependent, and the quantity of nutrient export with crop harvests. More clayey soils generally have a higher CEC than do sandy soils, which helps to retain base cations as they are released. In acidic soils, concentrations of Ca^{+2} and Mg^{+2} often remain elevated for decades after forest burning because these base cations replace H^{+} on the exchange complex. Continued annual burning helps release nutrients in remaining biomass back into the soil. However, intensive continual crop harvest, and the export of nutrients inside the crop, leads to a steady decline of these same nutrients from the soil in the absence of nutrient inputs. In particular, phosphate concentrations appear to decline steadily following forest conversion—due both to harvest export and to transformations in the soil that render phosphorus unavailable to plants—such that agricultural productivity is ultimately limited by this element (Garcia-Montiel 2000; McGrath et al. 2001). In addition, weeds become very aggressive in disturbed habitats, ultimately outcompeting crops for the steadily declining soil nutrients. Inevitably, without nutrient additions to offset export, weathered soils become nutrient depleted within two to three years of clearing, at which point a farmer must abandon the fields and clear more forest for continued production.

It is important to mention that slash-and-burn agriculture, also known as **shifting cultivation**, has been used for crop production around the world for centuries (Richter and Markewitz 2001). Key to its success in highly weathered soils is a long (15 to 20 years) fallow period during which the soil nutrient bank in abandoned fields recuperates through the regrowth of native plants and the reestablishment of nutrient-cycling processes. A study in Indonesia showed that shifting cultivation in sandy soils remained sustainable over a 200-year period because farmers practiced an adequately long fallow period, during which phosphorus availability was restored by **secondary forest** regrowth (Lawrence and Schlesinger 2001). However, as human population densities rise in forested areas, the fallow period is often shortened due to a lack of land. This inevitably leads to lowered agricultural productivity and more forest clearing.

PASTURES FOR LIVESTOCK

People throughout the world clear forests to raise livestock, often using slash-and-burn techniques, so many of the effects of conversion discussed above also apply to pastures. However, pasture creation also has a significant impact on soil physical properties, and many soil scientists believe that changes in soil physical properties due to poor land management are much more difficult to remediate than changes in soil chemistry (Sanchez et al. 1982). Research throughout the tropics has shown that forest conversion to pasture inevitably leads to an increase in soil bulk density due to compaction by heavy animals and the equipment used to maintain pasture (McGrath et al. 2001). As soils become compacted, they lose pore space, which critically influences soil water-holding capacity, aeration, and root penetration by plants, properties all very critical to the soil biota. Compaction also reduces soil infiltration by water, which can increase runoff in pastures and crop fields by as much as 28 times that on forested soils (Maass 1995). In ecosystems limited by precipitation, such as **tropical dry forests**, increased runoff also means less water available for plant uptake and thus lowered ecosystem productivity (Brady and Weil 2002).

Another major concern about forest-to-pasture conversion is loss of soil carbon storage, which results from increased rates of organic matter decomposition following clearing and cultivation (Houghton et al. 2000). Worldwide, organic matter in soils represents the third largest terrestrial store for carbon (Schlesinger 1997). Not only does burning forest biomass add CO_2 to the atmosphere, but higher rates of decomposition subsequently increase CO_2 efflux by respiring soil microbial populations. Thus, large-scale forest conversion to pasture, as is occurring in South America's Amazon Basin, may add to increasing atmospheric concentrations of CO_2, a greenhouse gas, by reducing soil carbon storage. To date, research in this area has rendered conflicting results. Studies comparing soils from forests and pastures of different ages and management intensities across the Amazon have shown higher, lower, and equal carbon contents between the two systems (McGrath et al. 2001). Pasture management, which largely controls pasture productivity, appears critical to maintaining soil carbon following conversion. One study suggests that soil carbon may actually increase in fertilized and seeded pastures due to rapid growth and death of pasture grass roots (Trumbore et al. 1995). However, similar to other developing regions, pasture management throughout the Amazon Basin is minimal and is unlikely to intensify (Kauffman et al. 1998), and thus the role of forest-to-pasture conversion in changing global soil carbon storage remains unclear.

POPULAR MISCONCEPTIONS ABOUT TROPICAL SOILS

Before ending our discussion on the effects of tropical forest conversion on soils, it is important to highlight some popular misconceptions about tropical soils and their management. These views have been long held due to popular press reports in the 1960s and qualitative observations of land use change in the tropics (Richter and Babbar 1991). For instance, early reports indicated that the conversion of forest to agriculture or other land uses resulted in the **laterization** or **desertification** of tropical areas (Goodland and Irwin 1975; McNeil 1966). Although bad land-management practices result in soil degradation in any **biome**—and there are plenty of misuses of soil in tropical areas—tropical soils do not always become biological deserts once managed.

Although 20% to 25% of the soils of the tropics are **oxisols** (see table 3.1), the most highly weathered soil order in the world, eleven of the twelve soil orders of the world are found in tropical areas (Sanchez and Logan 1992). Only **gelisols** (frozen soils) are not represented. Although approximately 40% of tropical soils are considered naturally low in fertility, 60% are considered moderately to highly fertile (Sanchez and Logan 1992). Thus, the idea that tropical soils are uniformly nutrient depleted and unmanageable is not true, because there is a wide range of soil diversity in the tropics, just as in other biomes. Thus, the response to forest conversion or other land use change requires a close look at each particular site and cannot be generalized.

In addition, we now know that tropical soils have chemical characteristics that are similar to soils in the temperate zone. For example, in studies looking at soil carbon, nitrogen, and organic matter, tropical soils were not significantly different from temperate zone soils for carbon, and some tropical soils maintain higher nitrogen and organic matter content compared to temperate soils (Buol et al. 1990). In fact, organic matter contents of mollisols (see table 3.1) in the midwestern United States are often not higher than those in tropical oxisols (Sanchez and Logan 1992).

It is important to point out that intensive land uses, especially in the more highly weathered soil orders, do require high quantities of outside inputs, such as fertilizers or green manures, to maintain plant yields over long time periods (Sanchez et al. 1982). Also, weed control in tropical soils is particularly difficult, and this affects crop and pasture viability soon after forest conversion. Soils that were relatively fertile when under forest cover can easily be degraded physically and chemically by human activities, but it is important to consider each site and soil type and not to generalize about "infertile tropical soils."

Throughout the tropics, research is under way to improve agricultural practices and to design systems that promote sustained low-input production in weathered soils. Much of this research highlights the critical role of soil organic matter and nutrient-cycling processes to the long-term maintenance of soil productivity. Examples of effective organic matter management techniques include fire-free field preparation, the use of nitrogen-fixing cover crops, and fallow fields enriched with useful native plant species (Lehmann et al. 2001; Schroth et al. 1999). Since the early 1980s, tree-based agroecosystems, often called **agroforestry** systems, have been under research because they maintain nutrient-cycling processes in a manner similar to natural forests. Planting trees as windbreaks or hedgerows, for example, can significantly decrease soil erosion. Often, trees grown as alley crops are pruned, and their cuttings are spread on annual crop rows as mulch that protects the soil and adds nutrients. In addition to conserving soil resources, agroforests frequently offer farmers benefits such as high-value cash crops, fuelwood, and ease of management compared to traditional cropping systems. While agroforestry systems do provide multiple benefits to landowners, any land use can become unsustainable if nutrient outputs with crop harvest are not offset by inputs (McGrath et al. 2000; Nykvist 2000).

Finally, the effects on soil fertility of forest conversion to agriculture depend upon the timescale considered. Although slash-and-burn agriculture may require a farmer to abandon land for 20 to 30 years, most studies demonstrate that soil productivity does return after an adequate fallow period, even after a century of shifting cultivation (Johnson et al. 2001; Nepstad et al. 2001). Moreover, secondary forests regrowing after agricultural abandonment eventually become mature forests that may not differ considerably from "pristine" **primary forests.** For example, a recent study using ground-penetrating radar demonstrated the existence of densely populated agricultural societies as late as 1600 AD in areas of the upper Amazon Basin previously presumed to be covered by pristine native forest (Heckenberger et al. 2003).

Current Research Themes and New Directions for Soils Research

We conclude our discussion on tropical soils and forest clearing by reviewing exciting topics in soils research, especially in the interdisciplinary field of biogeochemistry, which examines the cycling of chemical elements among the lithosphere, biosphere, atmosphere, and hydrosphere, especially in response to land use change. Although some of this work does not originate in the tropics, many of the key questions are related to understanding tropical soils and how ecosystems function in response to **disturbance**, such as land use change. For example, a concept still not well understood in soil is how soils develop over long timescales and how soil macronutrients (particularly nitrogen and phosphorus) enter and leave ecosystems over long time periods. One important set of studies of soil change over time has been conducted in the Hawaiian Islands under the guidance of Dr. Peter Vitousek of Stanford University. These studies examine how soils have evolved over a 4-million-year time period in a Hawaiian montane environment, and they are especially useful because these sites have similar climates, plant communities, and parent material. In fact, of the five soil-forming factors, only time since deposition is vastly different among the sites (Crews et al. 1995). Among other items, these studies found that losses of dissolved organic nitrogen and phosphorus were significant over time (Hedin et al. 2003) and that phosphorus persistently limited plant growth over long time periods (Vitousek and Farrington 1997).

Another series of studies that is currently changing the way we view soils has been conducted in Canada and Brazil in the BOREAS (Boreal Ecosystem-Atmosphere Study) and LBA (Large-Scale Biosphere-Atmosphere Experiment in Amazonia) projects, respectively. Both of these multidisciplinary studies link small-scale soil processes (soil nutrient and water dynamics, soil gas exchange, root dynamics) with larger-scale processes such as ecosystem-level gas exchange, carbon storage, and atmospheric responses to natural disturbances or land use change (Avissar et al. 2002; Margolis and Ryan 1997; Nobre et al. 2001; Sellers et al. 1997). The first field campaign of the BOREAS project was initiated in 1993, and the boreal forest in Canada was chosen as the study site in part because the soils of boreal forests contain almost half of the world's soil carbon stocks (Schlesinger 1997). In terms of soil science, the BOREAS project was particularly focused on CO_2 and CH_4 (methane) emissions from soils under different forest types and areas that were burned by forest fires (Amaral and Knowles 1997; Savage et al. 1997). The field-experiment stage of the LBA project started in 1998 and has resulted in over 100 scientific publications to date. Many of the LBA studies have examined the effects of agriculture, pasture conversion, or logging on Amazonian soil properties (Davidson et al. 2000; Schroth et al. 2002; Townsend et al. 2002) in an ongoing effort to understand the long-term implications of land use change in this important eco-region. These studies also will enable researchers to model soil nutrient processes to predict how land use will affect soil chemical status in the future.

In all biomes of the world, little is known about the belowground zone of influence of natural plant communities, such as whether belowground connections exist between different species, or what the significance is of plant-microbe-soil interactions. In particular, more information is needed on the size of root zones in natural plant communities and on how plant root structure is affected by soil texture and nutrient levels. Currently, experiments using nutrient analogs as tracers (cesium, lithium, rubidium, and strontium) are in use for the first time to measure root extension and nutrient uptake in natural soils (Casper et al. 2003). These results can then be used to develop models of belowground interactions among plants growing in natural communities.

Researchers are also seeking to determine if plants actually exchange resources belowground through shared associations of mycorrhizal fungi (Simard et al. 1997). For this type of study, researchers utilize carbon isotopes (^{13}C and ^{14}C) to detect belowground carbon transfers between different plant species. Other interactions between plants and the soil microbial community are largely unknown, and future studies that utilize molecular and stable-isotope techniques are called for not only to quantify the diverse population of soil microbes, but also to understand how different microbial groups affect plants in natural settings (Reynolds et al. 2003).

Another important question in future soils research deals with how parent material or bedrock is weathered. This is a key question because it allows scientists to estimate how fast the mineral nutrients that are locked away in rock are released into the soil. This is particularly important in regions with highly weathered soils that are experiencing intensive land use change (from natural forest to agriculture or plantation forestry). Because the measurement of weathering is very slow and difficult, strontium (Sr) isotopes are now used as an analog to calcium (Ca) to define calcium inputs and transport through soils (Bailey et al. 1996). Calcium is of particular interest because many researchers theorize that calcium losses will soon become a problem in the soils of the eastern United States (Huntington et al. 2000), especially in response to repeated logging of forests and plantations.

Finally, Richter and Markewitz (2001) called for the establishment of regional and worldwide networks of long-term soil experiments that should be formed to improve our understanding of soil and ecosystem change. This proposal calls for unification of the many ongoing research projects and institutions that conduct soils research, to improve the management of the world's forest, agricultural, and grassland systems that are reliant on the soil resource. These types of regional and global collaborations would ensure that soil changes created by large-scale land uses are well understood in all twelve of the world's soil orders.

Acknowledgments

The figures in this chapter were drawn by Samuel Grinstead, a forestry intern at the University of the South. Many of the definitions in the glossary were provided by the *Soil and Environmental Science Dictionary* published by the Canadian Society of Soil Science (Gregorich et al. 2001).

REFERENCES

Amaral, J. A., and R. Knowles. 1997. Localization of methane consumption and nitrification activities in some boreal forest soils and the stability of methane consumption on storage and disturbance. *Journal of Geophysical Research* 102:255–60.

Avissar, R., P. L. Silva-Dias, M. A. F. Silva-Dias, and C. Nobre. 2002. The Large-Scale Biosphere-Atmosphere Experiment in Amazonia (LBA): Insights and future research needs. *Journal of Geophysical Research-Atmosphere* 107(D20).

Bailey, S. W., J. W. Hornbeck, C. T. Driscoll, and H. E. Gaudette. 1996. Calcium inputs and transport in a base-poor forest ecosystem as interpreted by Sr isotopes. *Water Resources Research* 32:707–19.

Brady, N. C., and R. R. Weil. 2002. *The nature and property of soils.* 13th ed. Upper Saddle River, N.J.: Prentice Hall.

Bridges, E. M. 1997. *World soils.* 3rd ed. Cambridge, U.K.: Cambridge University Press.

Buol, S. W., P. A. Sanchez, J. M. Kimble, and S. B. Weed. 1990. Predicted impact of climate warming on soil properties and use. In *Impact of carbon dioxide trace gases and climate change on global agriculture.* Edited by B. A. Kimbal et al. ASA Special Publication 53. Madison, Wis.: Soil Science Society of America, 71–82.

Casper, B. B., H. J. Schenk, and R. B. Jackson. 2003. Defining a plant's belowground zone of influence. *Ecology* 84:2313–21.

Crews, T. E., K. Kitayama, J. H. Fownes, R. H. Riley, D. A. Herbert, D. Mueller-Dombois, and P. M. Vitousek. 1995. Changes in soil phosphorus fractions and ecosystem dynamics across a long chronosequence in Hawaii. *Ecology* 76:1407–24.

Cuevas, E. 1995. Biology of the below-ground system of tropical dry forests. In *Seasonally dry tropical forests.* Edited by S. H. Bullock et al. Cambridge, U.K.: Cambridge University Press, 362–83.

Davidson, E. A, L. V. Verchot, J. H. Cattanio, I. L. Ackerman, and J. E. M. Carvalho. 2000. Effects of soil water content on soil respiration in forests and cattle pastures of eastern Amazonia. *Biogeochemistry* 48:53–69.

Ewel, J., C. Berish, and B. Brown. 1981. Slash and burn impacts on a Costa Rican wet forest site. *Ecology* 62:816–29.

Garcia-Montiel, D. C., C. Neill, J. Melillo, S. Thomas, P. A. Steudler, and C. C. Cerri. 2000. Soil phosphorus transformations following forest clearing for pasture in the Brazilian Amazon. *Soil Science Society of America Journal* 64:1792–1804.

Goodland, R., and H. Irwin. 1975. *Amazon jungle: Green hell to red desert?* Amsterdam: Elsevier.

Gregorich, E. G., L. W. Turchenenk, M. R. Carter, and D. A. Angers, eds. 2001. *Soil and environmental science dictionary.* Canadian Society of Soil Science. Boca Raton, Fla.: CRC Press.

Heckenberger, M. J., A. Kuikuro, U. Tabata Kuikuro, J. C. Russell, M. Schmidt, C. Fausto, and B. Franchetto. 2003. Amazonia 1492: Pristine forest or cultural parkland? *Science* 301(5640):1710–14.

Hedin, L. O., P. M. Vitousek, and P. A. Matson. 2003. Nutrient losses over four million years of tropical forest development. *Ecology* 84:2231–55.

Houghton, R. A., D. L. Skole, C. A. Nobre, J. L. Hackler, K. T. Lawrence, and W. H. Chomentowski. 2000. Annual fluxes of carbon from deforestation and regrowth in the Brazilian Amazon. *Nature* 403(20):301–4.

Huntington, T. G. 2000. The potential for calcium depletion in forest ecosystems of southeastern United States: Review and analysis. *Global Biogeochemical Cycles* 14:623–38.

Johnson, C. M., A. H. Johnson, J. Frizano, I. C. Vieria, and D. J. Zarin. 2001. Carbon and nutrient storage in primary and secondary forests in eastern Amazonia. *Forest Ecology and Management* 147(2–3):245–52.

Juo, A. S. R., and A. Manu. 1996. Chemical dynamics in slash and burn agriculture. *Agriculture, Ecosystems and Environment* 58:49–60.

Kauffman, J. B., D. L. Cummings, and D. E. Ward. 1998. Fire in the Brazilian Amazon 2: Biomass, nutrient pools, and losses in cattle pastures. *Oecologia* 104:397–408.

Lal, R. 1990. *Soil erosion in the tropics: Principles and management.* New York: McGraw-Hill.

Lawrence, D., and W. H. Schlesinger. 2001. Changes in soil phosphorus during 200 years of shifting cultivation in Indonesia. *Ecology* 82:2769–80.

Lehmann, J., D. Gunther, M. Socorro da Mota, M. Pereira de Almeida, W. Zech, and K. Kaiser. 2001. Inorganic and organic soil phosphorus and sulfur pools in an Amazonian multistrata agroforestry system. *Agroforestry Systems* 53:113–24.

Maass, J. M. 1995. Conversion of tropical dry forest to pasture and agriculture. In *Seasonally dry tropical forests.* Edited by S. H. Bullock et al. Cambridge, U.K.: Cambridge University Press, 399–422.

Margolis, H., and M. Ryan. 1997. A physiological basis for biosphere-atmosphere interactions in the boreal forest: An overview. *Tree Physiology* 17:491–500.

Matson, P. A., P. M. Vitousek, J. J. Ewel, M. J. Mazzarino, and G. P. Robertson. 1987. Nitrogen transformations following tropical forest felling and burning on a volcanic soil. *Ecology* 68:491–502.

McGrath, D. A., M. L. Duryea, N. B. Comerford, and W. P. Cropper. 2000. Nitrogen and phosphorus cycling in an Amazonian agroforest eight years following forest conversion. *Ecological Applications* 10(6):1633–47.

McGrath, D. A., C. K. Smith, H. L. Gholz, and F. A. Oliveira. 2001. Effects of land-use change on soil nutrient dynamics in Amazônia. *Ecosystems* 4:625–45.

McNeil, M. 1966. Laterite soils. *Scientific American* 211:68–73.

Menaut, J. C., M. Lepage, and L. Abbadie. 1995. Savannas, woodlands and dry forests in Africa. In *Seasonally dry tropical forests.* Edited by S. H. Bullock. Cambridge, U.K.: Cambridge University Press, 64–92.

Neill, C., M. C. Piccolo, P. A. Steudler, J. M. Melillo, B. J. Feigl, and C. C. Cerri. 1999. Nitrogen dynamics in Amazon forest and pasture soils measured by ^{15}N dilution. *Soil Biology and Biochemistry* 31:567–72.

Nepstad, D. C., P. R. Moutinho, and D. Markewitz. 2001. The recovery of biomass, nutrient stocks, and deep soil functions in secondary forests. In *The biogeochemistry of the Amazon Basin.* Edited by M. E. McClain, R. L. Victoris, and J. E. Richey. New York: Oxford University Press.

Nobre, C. A., D. Wickland, and P. I. Kabat. 2001. The Large-Scale Biosphere-Atmosphere Experiment in Amazonia (LBA). *Global Change Newsletter* 45:2–4.

Nykvist, N. 2000. Tropical forest soils can suffer from a serous deficiency of calcium after logging. *Ambio* 29:310–13.

Perlin, J. 1989. *A forest journey: The role of wood in the development of civilization.* Cambridge, Mass.: Harvard University Press.

Reynolds, H. L., A. Packer, J. D. Bever, and K. Clay. 2003. Grassroots ecology: Plant-microbe-soil interactions as drivers of plant community structure and dynamics. *Ecology* 84:2281–91.

Richter, D., and L. I. Babbar. 1991. Soil diversity in the tropics. *Advances in Ecological Research* 21:315–89.

Richter, D. D., Jr., and D. Markewitz. 2001. *Understanding soil change: Soil sustainability over millennia, centuries, and decades.* Cambridge, U.K.: Cambridge University Press.

Sanchez, P. A., D. E. Bandy, J. H. Villachicia, and J. J. Nicholaides. 1982. Amazon Basin soils: Management for continuous crop production. *Science* 216:821–27.

Sanchez, P. A., and T. J. Logan. 1992. Myths and science about the chemistry and fertility of soils in the tropics. In *Myths and Science of Spoils of the Tropics.* Edited by R. Lal and P. Sanchez. Soil Science Society of America Special Publication 29. Madison, Wis.: Soil Society of America, 35–46.

Savage, K., T. R. Moore, and P. M. Crill. 1997. Methane and carbon dioxide exchanges between the atmosphere and the northern boreal forest soils. *Journal of Geophysical Research* 102:29, 279–29, 288.

Schlesinger, W. H. 1997. *Biogeochemistry: An analysis of global change.* San Diego: Academic Press.

Schroth, G., S. A. D'Angelo, W. G. Teixeira, D. Haag, and R. Lieberei. 2002. Conversion of secondary forest into agroforestry and monoculture plantations in Amazonia: Consequences for biomass, litter and soil carbon stocks after 7 years. *Forest Ecology and Management* 163:131–50.

Schroth, G., L. F. da Silva, R. Seixas, W. G. Teixeira, J. L. V. Macedo, and W. Zech. 1999. Subsoil accumulation of mineral nitrogen under polyculture and monocultural plantations and primary forest in a ferralitic Amazonian upland soil. *Agriculture, Ecosystems and Environment* 75:109–20.

Sellers, P. J., F. G. Hall, R. D. Kelly, A. Black, D. Baldocchi, J. Berry, M. Ryan, K. J. Ranson, P. M. Crill, D. P. Lettenmaier, H. Margolis, J. Cihlar, J. Newcomer, D. Fitzjarrald, P. G. Jarvis, S. T. Gower, D. Halliwell, D. Williams, B. Goodison, D. E. Wickland, and F. E. Guertin. 1997. BOREAS in 1997: Experiment overview, scientific results, and future directions. *Journal of Geophysical Research* 102:28, 731–28, 769.

Simard, S. W., D. A. Perry, M. D. Jones, D. D. Myrold, D. M. Durall, and R. Molina. 1997. Net transfer of carbon between ectomycorrhizal tree species in the field. *Nature* 388:579–82.

Sombroek, W. G. 1984. Soils of the Amazon Region. In *The Amazon: Limnology and landscape ecology of a mighty tropical river and its basin.* Edited by H. Sioli. Dordrecht, the Netherlands: D. W. Junk Publishers, 522–35.

Townsend, A. R., G. P. Asner, C. C. Cleveland, M. E. Lefer, and M. M. C. Bustamante. 2002. Unexpected changes in soil phosphorus dynamics along pasture chronosequences in the humid tropics. *Journal of Geophysical Research-Atmosphere* 107(D20):8067.

Trumbore, S. E., E. A. Davidson, P. Barbosa de Camargo, D. C. Nepstadt, and L. A. Martinelli. 1995. Below ground cycling of carbon in forests and pastures of eastern Amazonia. *Global Biogeochemical Cycles* 9:515–28.

Van Wambeke, A. 1992. *Soils of the tropics: Properties and appraisal.* New York: McGraw-Hill.

Vitousek, P. M., and H. Farrington. 1997. Nutrient limitation and soil development: Experimental test of biogeochemical theory. *Biogeochemistry* 37:63–75.

Wilson, E. O. 2002. *The future of life.* New York: Knopf.

SUGGESTED READINGS

Ewel, J., C. Berish, and B. Brown. 1981. Slash and burn impacts on a Costa Rican wet forest site. *Ecology* 62:816–29.

McGrath, D. A., M. L. Duryea, N. B. Comerford, and W. P. Cropper. 2000. Nitrogen and phosphorus cycling in an Amazonian agroforest eight years following forest conversion. *Ecological Applications* 10(6):1663–47.

McGrath, D. A., C. K. Smith, H. L. Gholz, and F. A. Oliveira. 2001. Effects of land-use change on soil nutrient dynamics in Amazônia. *Ecosystems* 4:625–45.

Sanchez, P. A., D. E. Bandy, J. H. Villachicia, and J. J. Nicholaides. 1982. Amazon Basin soils: Management for continuous crop production. *Science* 216:821–27.

HAPTER

4

From Farmers to Satellites

A HUMAN GEOGRAPHY ERSPECTIVE ON TROPICAL EFORESTATION

Peter Klepeis

Over the past three decades, concern about tropical **deforestation** has become widespread, with images of bulldozers, burning trees, and expanding agriculture and pastureland entering the collective imagination. Cries of alarm abound in both the scholarly and popular press: organisms that may hold the key to curing cancer are being snuffed out before science can uncover their secrets, the carbon once held in the forest is being released into the atmosphere and is contributing to human-induced climate change, indigenous cultures are being transformed by exposure to the outside world.

Along with concerns about the possible negative impacts of deforestation on society, there are many assumptions about what is causing the problem in the first place. Conventional wisdom pins the blame on population growth and, in particular, poor farmers who colonize the forests and practice **shifting cultivation** (**slash-and-burn agriculture**). Other explanations focus on logging or ranching activities. But most perceptions of tropical deforestation do not involve careful review of the evidence, leaving the complexity of the change dynamics hidden.

Understanding environmental change requires answering some basic, although difficult, questions about pattern and process. How widespread is the phenomenon? What is the magnitude of the transformation? In the case of tropical forests, is clearing occurring around the globe at equal rates and with similar patterns? And what are the processes involved? Why are people cutting down trees? What mix of political, economic, cultural, and biophysical factors leads people to use the land in the way they do?

The discipline of geography has a long tradition of studying environmental issues and linking patterns of environmental change to their underlying processes. Geographers contrast characteristics that are common to tropical deforestation worldwide with those that are unique to particular places. And they consider societal response. What should the global community do about one of the most important environmental issues of our time?

Why Worry about Tropical Deforestation?

Tropical deforestation is a problem that receives serious attention by many segments of society—scientists, the media, policymakers, and students, among others. Numerous concerns are raised (Brown and Pearce 1994; Myers 1992; Place 2001). Tropical forests contain extremely high **biotic diversity**. As sources of wood, food, and drugs, they provide subsistence to people and maintain economies from local to global scales, and they serve as the homes of unique cultures. Tropical forests are also sources and **sinks** (or reservoirs) of carbon, which makes them an important part of global climate systems. But beyond being beneficial to society either directly, in the case of food production, or indirectly, by helping to maintain a climate conducive to humans, ethical considerations point to the intrinsic right of the forest to exist regardless of its utility to people.

With such a broad range of ecological, economic, cultural, and ethical considerations, it is no wonder that a corresponding broad range of researchers from the natural sciences, social sciences, and humanities study tropical deforestation. Any environmental change involves a web of multifaceted dynamics. The integration of multiple perspectives, sources of data, methodologies, and explanatory frameworks is necessary, therefore, if the complexity of deforestation processes is to be uncovered. Part of this diverse team of investigators includes geographers.

Research Foci in Geography

WHAT IS GEOGRAPHY?

It is common for people to think of geography as a discipline that catalogs names and other information about places in the world. Certainly, description of the earth's cultural and physical features is part of what geographers do. The environments of particular places consist of social, biological, and physical conditions, and geographers often display this information using the maps everyone associates with them. In addition to characterizing patterns, however, geographers seek to explain why such patterns exist.

Textbooks introduce geography as explaining the "why of where"; the assumption that location matters in natural and social processes underpins most analysis (National Research Council 1997). Why are tropical forests located where they are? What combination of factors produces forests in these specific locations? Why are they so rich in biotic diversity? Why is deforestation occurring where it is? What are the processes of environmental change affecting a given location?

Geographers are united by their efforts to connect landscape change with the multiple social and biophysical factors involved; in this way, geography represents, in a sense, a disciplinary middle ground. If geography were a three-legged creature, it would position itself with a foot in each of the broad categories of intellectual inquiry: the natural sciences, the social sciences, and the humanities.

Geography occupies a disciplinary middle ground in part because of its place-based focus. The **human–environment conditions** (the combination of social and biophysical characteristics) of a particular place are linked to overlapping factors (see figure 4.1). Sack (1997) argues that the ways in which people relate to their natural environment, and thus the ways they choose to manage natural resources, depend on forces in nature; social relations (the ways in which people interact with each other); and meaning, that which—apart from thoughts that are generated from our interaction with other people—is produced from within each of us.

This framework may seem abstract at first; however, think of how these forces shape the ways in which land management decisions are made. Natural forces and conditions place constraints on how humans behave. An area with more fertile soils and adequate precipitation facilitates agricultural production, whereas a dry zone with less-rich soils is a more challenging context for growing food. Clearly, aspects of how individuals live are also dependent on social relations, such as the rules and laws by which

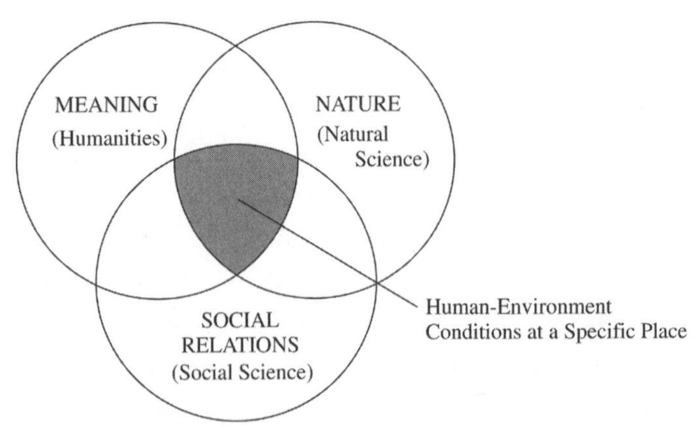

Figure 4.1. A central goal of research in geography is explaining environmental changes tied to specific places through interdisciplinary connections (after Sack 1997).

society is governed. Whose decision is it, for example, to cultivate a parcel of land or not—a government official, the head of household, or the collective community? Finally, the actions of individuals also depend on their worldviews, their notions of morality, and their sense of the aesthetic—the realm of meaning. Is there a spiritual connection to the forest for some people? Or do individuals hold the view that nature is there to serve the interests of human beings and is therefore to be exploited to the utmost? In short, geographers try to uncover how biophysical processes, social processes, and individual conceptualizations of nature intersect at a given place to cause a particular set of human-environment conditions.

PATTERNS

The world's tropical forests have undergone rapid transformation in recent decades, with the Food and Agriculture Organization (FAO) estimating an 8.7% (152 million hectare) reduction of tropical forest area worldwide between 1990 and 2000 (FAO 2001). Achard and colleagues (2002) estimate the global annual deforestation rate for humid tropical forests between 1990 and 1997 to be at 0.52%, although this rate drops to 0.43% if you consider net forest change—that is, both cutting and regrowth. In their analysis of 108 published studies of tropical deforestation worldwide, Geist and Lambin (2001, 76–78) put these figures in perspective by characterizing relatively low (0.3% to 0.7%) and relatively high (1.9% to 2.9%) mean annual deforestation rates by region. Rates do change through time. Kummer (1992) finds that deforestation rates in the Philippines between 1947 and 1987 range from 1.74% per year to upwards of 3.64% per year, with an overall loss of forest estimated at approximately 84,900 square kilometers. Cortina and colleagues (1999) find that between 1975 and 1981, the annual rate of deforestation in the southern Yucatán Peninsula was an extremely high 5.2% per year but dropped to 1.4 % per year between the mid-1980s and 1990. The average rate of deforestation between 1969 and 1997 for the same general region, however, was a relatively low 0.4% per year (Turner et al. 2001). But while improvements continue to be made, assessments of forest cover change remain crude, flawed in part by inadequate monitoring, inconsistent definitions of what constitutes deforestation, and poor data quality (Fairhead and Leach 1998; Matthews 2001). There is consensus that tropical deforestation is a serious problem and that the scale of human impact is great; however, our understanding of where and when forest disturbance has occurred is incomplete.

There are at least four reasons why determining deforestation rates across time and space is challenging. First, the extent of tropical forests worldwide is large, covering some 1.74 billion hectares (FAO 2001). It is difficult to know what kinds of impacts are occurring across such a vast expanse, in particular given the geographical isolation of many forests. Ground-level surveys are helpful—if carried out effectively—but they only provide documentation on human impact for select locations. Second, the cost of mounting comprehensive ground surveys may be prohibitive, making continuous monitoring difficult, especially given that countries with tropical forests are often poor. Third, there is debate over how to define what is meant by tropical deforestation. If a specific unit area of forest has 90% of its canopy cleared (the FAO's definition), is that area considered "deforested"? Is an area that is cleared of forest one year but that experiences regrowth in subsequent years still defined as deforested? What about other kinds of human impacts on forests that don't involve clear-cutting, such as selective logging or partial cuts? It is difficult to compare estimates of forest disturbance, especially those by different researchers, without consistent answers to these questions. Finally, inadequate or unknown baseline data about the condition of forests long ago, before significant human occupation of them, limits what researchers can say about overall human impact on forests.

The analysis of **remotely sensed imagery** (information from satellites or aerial photographs) using **geographic information systems** (GIS) addresses some of these challenges and enhances the assessment and monitoring of deforestation patterns. A GIS is a computer-based technology that allows researchers to analyze, manage, and display spatial information, such as satellite imagery or road networks. Imagine a stack of paper maps each representing a different theme: political boundaries; soil characteristics; population density; precipitation patterns; **land cover** (e.g., cleared land, forest, grassland, wetland); and other phenomena that vary across space. A GIS allows the researcher to manage digital versions of these data layers and to analyze the importance of each for a particular environmental issue.

The use of satellite imagery and GIS for monitoring deforestation allows the vast expanse of tropical forests to be monitored much more cheaply in comparison to ground-level surveys. And by analyzing imagery from multiple periods over the past three decades (satellite imagery became available for research in the mid-1970s), investigators can also identify the rate of deforestation (area cleared per year) and patterns of forest disturbance through time. In addition, GIS and remote sensing help uncover the dynamics of deforestation. A simple and intuitive example is the finding that where there are roads and human settlements, there tends to be more forest disturbance than in isolated and thinly populated areas. But less obvious connections become apparent as well; **topography** (slope, aspect, elevation) or farm size, for example—variables that may influence the viability of agriculture—are often important in determining whether settlers choose to clear and cultivate a given location.

PROCESSES

GIS and remote sensing, by their very nature, involve analysis of spatially explicit and quantifiable data that are mapped easily onto specific locations. But there are other factors driving deforestation that are qualitative and not spatially explicit—that is, they cannot be tied to specific geographic coordinates. To fully answer why and how tropical deforestation occurs, therefore, requires additional approaches, such as surveying local land managers to understand their decision-making processes, exposing the history of development policy through archival analysis, and directly observing regional land use systems.

There is significant debate over the degree to which variables at the local or community scale (e.g., population growth) drive most of the deforestation, versus factors that manifest themselves at regional, national, or international scales, such as government policies or demand for commodities produced in the tropics (e.g., coffee, timber, rubber). Put another way, mass media often blame tropical deforestation on farmers who use shifting cultivation to feed their families, while **nongovernmental organizations** (NGOs) and environmental activists tend to blame government policies or logging companies. Who is right?

It is clear from this question that knowing the why, where, and how of tropical deforestation is more than an interesting environmental problem. The answer influences how society chooses to respond. If shifting cultivation is the main culprit, should alternative livelihood opportunities be made available to forest inhabitants? Can forest cover and biotic diversity be maintained while also providing income and subsistence for local people?

The kinds of answers to the "what is causing tropical deforestation?" question, minimally, can be split into two broad kinds (Klepeis 2003). The first focuses on **human agency**, or the ability of individuals and small groups to manipulate conditions to their own benefit. In this form of explanation, the immediate agents of change (i.e., local-scale land managers and inhabitants) are identified as the primary players driving deforestation. The assumption is that regional population growth puts increasing pressure on human-environment conditions, leading inhabitants to open up frontier forests to expand their livelihood opportunities. A second form of explanation focuses on **social structures**, or the rules, conventions, and restraints that govern individual human behavior. In this case, political and economic conditions (defined by a mix of social structures) are shown to empower certain economic agents (e.g., logging or mining companies) that clear the forest with little regard for the social and ecological consequences, but at the same time to disempower the forest inhabitants and their capacity to manage the forest sustainably. Increasingly, it is recognized that both of these forms of explanation—structure and agency—reflect the human-environment conditions causing deforestation around the world.

Studies of the role of human agency in deforestation tend to focus on local- or proximate-scale dynamics. Throughout history, time and again, people have shown themselves remarkably adaptive to shifting human-environment conditions. Part of adaptation involves transforming the environment in ways that suit societal needs. For example, from Australia to Africa to the Americas, and in a broad range of biophysical contexts, indigenous groups have used fire to create advantageous conditions for hunting and agriculture. Whatever the mix of flora and fauna and natural resources available in a particular location, humans have shown themselves capable of exploiting the environment to their advantage. In tropical forests, an example of this effective adaptation is shifting cultivation, a system that is well suited to the ecology of the forest and that represents an efficient mode of food production (see below). But in the context of human-environment conditions that change too fast, such as the 20th-century boom in population growth and frontier settlement, the capacity for adaptation is weakened: short-term livelihood needs can displace long-term, more sustainable land management strategies. And, if no alternative livelihood opportunities are available, the result usually includes clearing more forest.

Analysis of social structures shifts the focus away from just the local scale and considers how regional inhabitants may be forced to clear forests due to factors beyond their control, emanating from outside their sphere of influence (Folke et al. 1998). Two important examples are the influence of government policies and markets. Government policies affect individual behavior by making some activities illegal and by encouraging others deemed advantageous to society, and by segregating access to resources for different groups of people (e.g., taking away indigenous land and leasing it to logging companies or setting it aside as a nature reserve). Federal policies that encourage forest clearing through tax incentives; that provide roads allowing access to frontier locations; or that invest in particular land uses, such as large-scale ranching initiatives, take agency away from local inhabitants. And fluctuations in market demand for particular forest commodities, such as rubber, can negatively affect local livelihoods, forcing inhabitants to search for alternative land uses. In essence, explaining deforestation in terms of social structures asks you to consider local-scale dynamics as more than a closed system, and to follow the chain of linkages away from where forest clearing occurs to the national and international markets and the federal development policies that underlie the behavior of local land managers.

In reference to the widely assumed link between population growth and environmental degradation, for example, analysis of social structures shows that population growth in an area undergoing deforestation is often due to social dynamics elsewhere. In many cases, an influx of people from outside the region, rather than natural population growth (as births exceed deaths), explains the population change. Land scarcity or lack of employment opportunities in urban areas may cause the federal government to use colonization programs in forest frontiers as a release valve to diffuse the political tension created by widespread poverty and unemployment. The Brazilian government moved people from coastal regions into the Amazon basin in the 1970s and 1980s to provide opportunities for disenfranchised people as well as to intensify the exploitation of the country's natural resources in the hopes of fostering economic development (Hecht 1985).

Current Research in the Field

Research on the patterns and processes of tropical deforestation has led to important insights that hold implications for possible societal responses to the problem. From this large body of work, four themes are emphasized: 1) perceptions of shifting cultivation, 2) forest regeneration or secondary growth, 3) boom-bust cycles of deforestation, and 4) strategies to combine forest conservation with economic development. These themes are representative of research in multiple contexts across the world, although much of the evidence referred to here is from Latin America. A case from southeastern Mexico is used here to exemplify the crosscutting themes in deforestation research and efforts to integrate various forms of explanation in a single research project (see box 4.1).

Box 4.1. The SYPR Project: Linking Pattern to Process

The land cover and land use change—The Southern Yucatán Peninsular Region (SYPR) project is a collaborative research venture between Clark University, Harvard Forest (Harvard University), and El Colegio de la Frontera Sur, a research institute in Mexico. The 22,000-square-kilometer study region is in southeastern Mexico, abutting the borders of Guatemala and Belize. It is part of the largest contiguous tropical forest remaining in Mexico and Central America. The region underwent widespread deforestation in the distant past (by the Classic Maya civilization, AD 300–900), and, since the 1960s, there have been high rates of deforestation due to extensive road building, colonization and land grants, large agricultural projects, and expanding small-scale agriculture (Turner et al. 2004). In seeking to explain both the patterns and the processes of deforestation, the SYPR project links two research approaches: in-depth ground-level studies that focus on case studies, detailed assessments of human-environment conditions, and social and biophysical change processes; and the use of satellite imagery and aggregate socioeconomic data to capture broad patterns of forest disturbance.

Patterns of deforestation—Between 1969 and 1997, the total **anthropogenic disturbance** (that caused by humans) approached 10%. The region underwent a boom in forest clearing in the 1970s and early 1980s due to government-directed colonization and large agriculture projects for rice and cattle in lowland areas, as well as government-subsidized clearing for shifting cultivation in upland forests. There was a lull in forest clearing after these large projects failed and government investment dried up, only to be followed by a subsequent increase in deforestation rates in the face of spontaneous colonization by settlers from land-scarce regions in other parts of Mexico. But, over the past 10 years, now that most land in the area has been allocated, rates have again slowed. Much of the forest that was once cleared has reverted to multiple stages of secondary growth. For the 1995–2001 period, satellite imagery shows less land in cultivation than in succession or fallow and a reduction in the cutting of mature forest.

Processes of deforestation—20th-century trends in deforestation are primarily linked to structural shocks and, until recently, not the agency of local farmers. Government policies opened up the region to international forest companies

(continued)

Box 4.1. (*Continued*)

in the first half of the 20th century, leading to the construction of road networks that enabled subsequent settlement. In the 1970s and 1980s, policies targeted the area for settlement and agricultural development. One of the reasons for the focus on agriculture and not on the selective logging of mahogany and the extraction of chicle (a tree resin that serves as a base for chewing gum) was the prior depletion of mahogany stands due to shortsighted management and the bust of the chicle market after a synthetic chewing gum base was created after World War II (activities surrounding chicle exploitation led to minimal forest disturbance). Following the bust in forestry, the failure of the large-scale agriculture and ranching initiatives, and the decline in land available for new settlers, deforestation rates slowed.

Perceptions of human impact on the forest—The species composition of mature forests today bears the mark of Classic Maya activity from over 1,000 years ago and from selective logging from the mid-20th century. Recognition of the human element in the forest's history may partly explain why most conservation initiatives in the region now recognize the need for an integrated land management strategy that embraces both sustainable economic uses of the forest and nature reserves.

National policy responses—Government attempts to slow deforestation are mixed at best but include the creation of the Calakmul Biosphere Reserve in 1989, incentives to encourage more intensive agriculture (to maintain or raise food production without clearing new land), and investment in an ecological and archeological tourism scheme. The more likely factors in the lower rates, however, are the lack of federal investment in new projects that involve major clearing; the stabilization in regional population; farmer-led experiments with new agricultural systems (e.g., jalapeño chili production); and the rise of community forestry projects.

Note: Major funding for the SYPR project (1997–2000) was from NASA's LCLUC program and the Center for the Integrated Studies of the Human Dimensions of Global Change (Carnegie Mellon University, NSF).

PERCEPTIONS AND ADAPTATION

Three important findings about tropical forests and their inhabitants are calling into question some long-standing assumptions: 1) traditional food production systems, such as shifting cultivation, *are* often well adapted to human-environment conditions; 2) assumptions that tropical forests have an intrinsically low **carrying capacity**—a term that notes limits on the capacity of the environment to support a growing population—ignore the many historical examples of intensive agriculture in tropical forest locations where high population densities were supported; and 3) tropical forests are *not* "virgin" or "pristine" but instead contain a human imprint both from ancient and more recent activities.

First, despite conventional wisdom that shifting cultivation is a maladaptive system responsible for degrading forests, this form of agriculture is energy and labor efficient and is well adapted to a tropical forest ecology, mimicking the forest's complex structure and high biotic diversity (Denevan 2001, 83–90; Rappaport 1971). Shifting cultivation involves cutting the vegetation for a given plot of land, allowing it to dry, and burning it to both clear the land of vegetation and to enrich the soil with ash. Traditionally, after a few years, as soil fertility declines and weeds become a problem, the land is allowed to revert to forest over the course of many years, a phase in which the land is in fallow. In the meantime, the farmer repeats the process in numerous other locations and then returns to the original spot after 10 to 20 years. In areas with low population density, where forested land is plentiful, the system can sustain production levels through time without disrupting the forest's basic structure and composition. Indeed, Rappaport (1971) finds a lesson in shifting cultivation for advancing sustainable agriculture today, in that it maintains diverse, complex agroecosystems that are more resistant to pests than are the monocropped fields of modern agriculture. There is a cautionary tale, however. As population densities increase and land becomes scarcer, the length of the forest fallow period decreases, which may lower biotic diversity, crop yields, and the forest's ability to recover.

This potential for declining agricultural productivity in the face of increasing population and land pressure has led many researchers to believe that tropical forests have a low carrying capacity. The argument is that most tropical environments are not suited for intensive occupation and land use (e.g., because of pests, thin soils, and high humidity that make food storage difficult), and thus human population levels must remain small. But carrying capacity is determined largely by technologies and land management strategies rather than by environmental conditions (Brush 1975).

The southern Yucatán Peninsula in southeastern Mexico provides a case in point (Turner 1990). Current population densities in the region are quite low, averaging only a few people per square kilometer. In contrast, average population densities during the Classic Maya period (AD 300–900) approached 100 people per square kilometer, with those in urban areas exceeding 1,000 people per square kilometer. To maintain such high population densities, the Classic Maya civilization—which was spread throughout the tropical lowlands of southern Mexico, Guatemala, Belize, Honduras, Nicaragua, and El Salvador—used intensive and complex agricultural systems. It did not rely solely on shifting cultivation. The lesson is that, while environmental conditions in tropical forest locations may place constraints on agricultural production, technical sophistication and adaptive resource management can raise the capacity for food production. The concept of carrying capacity is, therefore, best used with caution; it is helpful to the degree that it suggests limits to the natural resource base (e.g., topsoil, groundwater, biotic diversity), but it should not be used to indicate some absolute number of people that nature can support.

Finally, there is a pervasive assumption that tropical forests were sparsely occupied before the 20th century and that significant human impact on the forests is a recent phenomenon. Denevan (1992) calls this assumption the **pristine myth**. But, in contrast to notions that precolonial forests were untouched by people and were pristine, humans have a long history of transforming tropical forest landscapes in many parts of the world (Denevan 2001; Whitmore and Turner 2001). The legacy of this impact is seen in forest structure and species composition; in archeological evidence of intensive agricultural systems (e.g., terraces and raised fields); and in the evidence of anthropogenic soils (deep, rich, fertile soils created by people practicing intensive agriculture) (Denevan 2001). For example, one estimate is that *terra preta*—"black earth," a very fertile soil created by indigenous groups long ago—covers 10% of Amazonia (Mann 2002).

Returning to the southern Yucatán Peninsula example, the Classic Maya civilization had almost completely deforested the region by AD 900–1000, after which it collapsed due to a combination of economic, political, and environmental factors (Turner et al. 2003). Subsequently, human impact on the land became miniscule, and mature tropical forests returned over the course of hundreds of years. In other words, the forest we see today is essentially secondary growth. And its structure and species composition bear the human imprint of the Classic Maya (e.g., Maya orchard gardens built up economically useful species that are still found today).

Knowing forest histories, such as that of the Classic Maya region, is important. Our assumptions about whether or not the forest is pristine affects how we choose to manage it today. If places once thought to have existed for millennia as primordial forest are found to be human creations, at least in part, then arguments to preserve them, away from people, become problematic. If the very nature that we celebrate as "wilderness" carries a human imprint, then why not see people as part of nature? This is not to say that we should not worry about deforestation and should permit widespread clearing; however, it does suggest that we should be able to figure out a way for people to coexist with forests (see section on "societal response" below).

FOREST REGROWTH AND FISH BONES

"Forests are not static, but living entities; they regenerate and grow in favorable environmental conditions. They can be replanted as well as removed, managed as well as neglected, and their productivity hindered or encouraged" (Williams 1990, 195).

The degree of human impact on forests varies across space and time. The spatial pattern of forest disturbance captured in remotely sensed imagery provides insights into processes of land use change, such as the role of directed colonization schemes. But imagery most effectively detects forested and nonforested areas, simple categories that hide the complexity of heterogeneous landscapes. Human use and natural perturbations disturb forest and lead to multiple stages of succession, from very young to mature regrowth. These patches of secondary growth are often the product of shifting cultivation, which lets forest shift in and out of use over time; land abandonment; or biophysical shocks such as fires and hurricanes. Two other land use systems that make for a diverse landscape are **agroforestry**, a form of land management that incorporates a mix of trees and crops, and **forest plantations**, which are cultivated forests that lack species diversity but that represent areas of maintained forest cover nonetheless. Explanations of the patterns of deforestation should reflect the dynamic nature of forests and their use.

The complexity of land uses is not always easy to detect in a satellite image, which is why researchers are often forced to rely on forest-nonforest designations. Despite this limitation, the patterns of forest clearing captured by remotely sensed imagery suggest underlying processes of change. A widely recognized symbol of tropical deforestation is the fish-bone pattern (see figure 4.2). While explanations of the pattern are fairly simple, the example underscores the importance of integrating analysis of remotely sensed imagery with intensive ground-level studies—the practice of linking pattern to process. A common scenario has the federal govern-

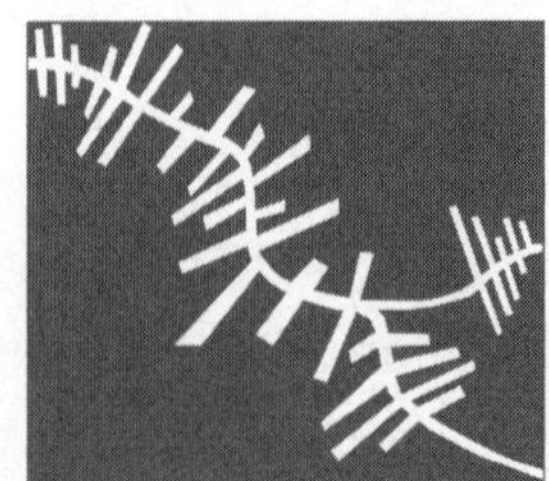
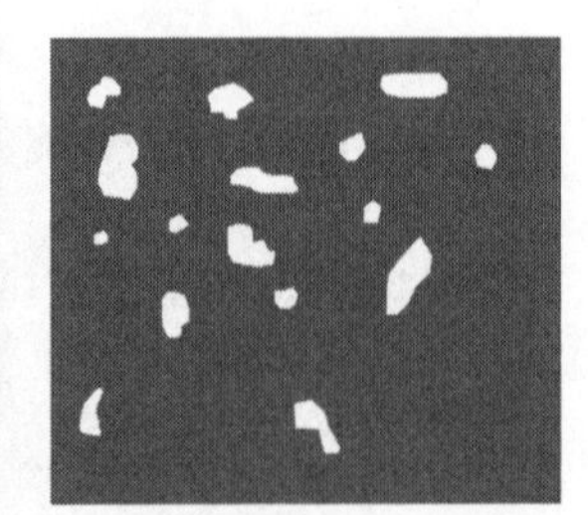

Figure 4.2. Three forest-nonforest spatial patterns (after Geist and Lambin 2001: 66).

ment of a developing country encouraging the settlement of frontier forests to exploit land resources and to foster economic development. A road is built that connects urban centers with the region to be settled. The government explains to settlers that they will receive title to those lands that are cleared of trees because it seeks to promote agricultural production for the market. Settlers arrive via the road and gain access to parcels of land with rectangular shapes. People tend to clear trees beginning close to the road and then gradually proceed away from it. This scenario explains why remotely sensed imagery shows rectilinear clearings adjacent to the road. It also shows how explaining the patterns of deforestation seen in imagery requires understanding the underlying factors in forest clearing, such as government policy. Two other examples of patterns found in imagery are geometric patterns, which reflect large-scale clearing activities such as cattle ranching, and diffuse patterns, which reflect smallholder agricultural and usually shifting cultivation.

The study of patterns of deforestation and the phenomenon of forest regeneration are closely linked. For example, forest recovery (regrowth) in temperate areas of the world tends to occur on peripheral lands, away from roads. This occurs in most cases because land that was once used for agriculture is abandoned due to increased off-farm labor opportunities in more urban areas. Rudel, Bates, and Machinguiashi (2002) find, however, that in the Ecuadorian Amazon, regrowth is occurring close to roads. Farmers with small landholdings are abandoning cattle ranching, which often occurred close to the road, to practice shifting cultivation. The reason is that markets have changed such that agricultural produce for urban markets is in greater demand than livestock. The farmers have an incentive, therefore, to use their lands for agriculture as opposed to ranching, which necessitates changing the land use system and allowing much of the land near the road to revert to forest fallow (unless chemical fertilizers and pesticides are readily available, in which case the land near the road may be used for agriculture right away). Presumably, this secondary growth will be used for shifting cultivation as times goes by.

Despite assumptions that deforestation inevitably leads to devastation, such as permanent conversion from forest to pasture, in some areas of the tropics, rates of forest recovery are becoming higher than the rates of ongoing forest loss, resulting in a net gain of forest cover (Perz and Skole 2003). It is true that secondary forests are not the same as mature ones, but they are not ecologically trivial. Under the right human-environment conditions and in as little as three decades, biomass, canopy height, and even species composition can start to approach precut levels (Finegan and Nasi 2004). The recovery of sections of once deforested areas underscores how forest change is not a linear process. Moran and Brondizio (1998) find that many cut or burned areas experience rapid regrowth and that these patterns of regrowth can be linked to differences in soil fertility. Skole and colleagues (1994, 316) show that 42% of new agricultural land created between 1988 and 1989 in the Brazilian state of Rondônia came from clearing secondary growth. Recent work on deforestation in the southern Yucatán Peninsula (see box 4.1) reinforces the importance of understanding how farmers and other land managers use secondary forest. Significant tracts of mature forest were cut in the 1970s and 1980s; however, in the 1990s, farmers tended to clear successional growth (Turner et al. 2001). Research into why this is so holds implications for how more sustainable uses of the forest might include use of secondary forest as opposed to more mature vegetation.

These examples demonstrate the variability in the extent of deforestation and the need to specify carefully the spatial and temporal scale of analysis when discussing human impacts on forests. Rates fluctuate across space and time due to variability in the forces driving the deforestation. Shifts in the demand for forest products, such as in the Ecuadorian case; changes in government policy (e.g., the implementation or cancellation of a colonization program); or increased pressure to conserve **biodiversity** may factor into whether there is more or less forest clearing in any given historical period.

BOOM-BUST CYCLES

Fluctuating rates of deforestation through time reflect shock and intershock periods. Shocks often manifest themselves as sudden shifts in government policy and commodity prices (Kummer and Turner 1994). The classic example of a shock is the construction of a new road into a previously unoccupied or thinly occupied area. Almost overnight, an inaccessible forest is opened up to newcomers, whether they be loggers, miners, ranchers, or subsistence farmers. Likewise, a shift in market demand for beef, timber, grain, minerals, or some other resource that might be extracted from the land may lead to new interest in opening forested regions for economic use. But shocks do not always lead to increased forest disturbance. The rise in international conservationism in the 1980s and 1990s led many governments to establish nature reserves. The effectiveness of these conservation strategies is mixed; however, in many instances, reserves led to reduced deforestation rates for select locations.

Deforestation is also tied to forces operating between the shock periods, such as factors involving the agency of local people. Coomes and Barham (1997) show that forest inhabitants in Peru frequently change their livelihood strategies over the course of their lifetime depending on access to land, labor, and capital. Land management is also explained by analyzing intrahousehold dynamics, such as differences in the roles of men and women in managing forest resources, or the age structure of household occupants. Sadoulet, de Janvry, and Davis (2001) find that the degree to which Mexican farmers intensify agricultural production (and reduce reliance on extensive shifting cultivation) depends largely on their access to capital with which to invest in production, to extension services that introduce new land management strategies, and to local organizations that allow farmers to make linkages to markets outside the region.

SOCIETAL RESPONSE

Linking pattern to process can be contentious, as demonstrated in a recent debate in the journal *Science* about the causes of deforestation in Amazonia. One group of researchers (Laurance et al. 2001, 2005) identifies highways and roads as the primary factor driving deforestation. While acknowledging the importance of road infrastructure, other researchers (e.g., Bruna and Kainer 2005; Schaeffer and Rodrigues 2005) express concern over the implication of what they see as an overly simplistic finding: slow road creation and you'll slow deforestation rates. They call for a recognition of both the broad suite of underlying forces of Amazonian development, and therefore the acknowledgment that multifaceted societal responses are necessary to address the deforestation problem, and the role that market access (via roads) has in achieving a balance between environment and development in the region.

In considering the possibility of balancing environment and development goals, scientists, politicians, and the general public invariably rely on the concept of **sustainable development**, the notion that economic growth and material development can occur at the same time as social welfare, equity, and environmental conditions are maintained or improved (Brundtland 1987). But is sustainable use of tropical forest resources possible? The assumption of those who promote sustainable development is that local people must be able to earn a good living from tropical forests if they are to have the incentive to stop clearing it. The use of **extractive reserves** is one popular idea to this end.

Extractive reserves are tracts of forest in which forest clearing is prohibited. Local people are allowed to enter the reserve to extract forest products, such as rubber, chicle, or nuts, or to selectively log. These forest commodities are then sold in regional, national, or international markets to support local livelihoods and to offer alternatives to uses that require clearing the forest.

In the short term, if demand for extractive forest commodities is high, extractive reserves work fairly well. Analysis over the long term paints a different picture, however. Markets in forest commodities are not stable, fluctuating dramatically over time (Coomes 1995). In the cases of chicle and rubber, synthetic substitutes were created that caused demand for these forest products to plummet. What are locals to do if their source of income disappears? They search for other strategies to make ends meet, such as agriculture, ranching, or off-farm labor.

But if extractive reserves cannot prevent deforestation from occurring, then what should be society's response? The answer you get depends on the person you ask. Researchers in geography and conservation biology increasingly call for diversified land management systems that integrate the use of nature reserves with sustainable economic uses. In other words, protected areas must be tied to poverty alleviation. Some areas are so ecologically sensitive that biotic diversity and other **ecological services** cannot be subject to human use without degradation occurring. But understanding of how to use the land sustainably is improving. Integrating extractive industries with intensive agriculture, agroforestry, and other productive land uses, along with the provision of livelihood strategies not necessarily connected to the land, decreases the vulnerability of local people to poverty and reduces the prospect that they will be forced to cut the forest unsustainably (Browder 2001). If diversified, the capacity of the regional economy to absorb fluctuations in commodity prices or other perturbations is enhanced, which is good for both the people and the environment. How to provide this diversity is subject to considerable debate. However, researchers increasingly recognize the need for democratic decision making; "active, inclusive, and iterative communication" between regional stakeholders tends to lead to more informed decisions, more sustainable use systems, and a better balance between environment and development (Cash et al. 2003, 8088).

Conclusion

Research in geography shows clearly that understanding tropical deforestation dynamics necessitates making local to global, cross-scalar connections. Deforestation is caused by a web of socioeconomic, political, and biophysical factors that manifest themselves at intrahousehold, parcel, community, regional, national, and international scales. But by comparing case studies and context-specific deforestation dynamics with global patterns and trends, geographers are starting to navigate their way through this complexity and to improve our understanding of human-environment relationships. By exposing the range of factors that affect local production systems, geographers are participating in a larger discourse on how to achieve sustainable development. By debunking the myths of simplistic, linear causal explanatory chains, research findings underscore the lack of a silver-bullet solution to the deforestation problem. What is left is the challenging process of designing diverse landscapes that simultaneously fit human and **ecosystem** needs.

REFERENCES

Achard, F., H. Eva, H. J. Stibig, P. Mayaux, J. Gallego, T. Richards, and J. P. Malingreau. 2002. Determination of deforestation rates of the world's humid tropical forests. *Science* 297:999–1002.

Browder, J. O. 2001. Alternative rainforest uses. In *Tropical rainforests: Latin American nature and society in transition.* Edited by S. Place. Jaguar Books on Latin America, no. 2. Wilmington, Del.: Scholarly Resources, 196–205.

Brown, K., and D. W. Pearce., eds. 1994. *The causes of tropical deforestation: The economic and statistical analysis of factors giving rise to the loss of the tropical forests.* London: University College London.

Bruna, E. M., and K. A. Kainer. 2005. A delicate balance in Amazonia. *Science* 307:1044–55.

Brundtland, G. H. 1987. *Our common future: World Commission on Environment and Development.* Oxford: Oxford University Press.

Brush, S. B. 1975. The concept of carrying capacity for systems of shifting cultivation. *American Anthropologist* 77:799–811.

Cash, D. W., W. C. Clark, F. Alcock, N. M. Dickson, N. Eckley, D. H. Guston, J. Jager, and R. B. Mitchell. 2003. Knowledge systems for sustainable development. *Proceedings of the National Academy of Sciences* 100(14):8086–8091.

Coomes, O. T. 1995. A century of rain forest use in western Amazonia. *Forest Conservation and History* 39(3):108–20.

Coomes, O. T., and B. L. Barham. 1997. Rain forest extraction and conservation in Amazonia. *Geographical Journal* 163(2):180–88.

Cortina, V. S., P. Macario Mendoza, and Y. Ogneva-Himmelberger. 1999. Cambios en el uso del suelo y deforestacion en el sur de los estados de Campeche y Quintana Roo, Mexico. *Boletin del Instituto de Geografia de la UNAM* 38:41–56.

Denevan, W. M. 2001. *Cultivated landscapes of native Amazonia and the Andes.* Oxford: Oxford University Press.

———. 1992. The pristine myth: The landscapes of the Americas in 1492. *Annals of the Association of American Geographers* 82:369–85.

Fairhead, J., and M. Leach. 1998. *Reframing deforestation: Global analyses and local realities.* New York: Routledge.

FAO (Food and Agriculture Organization of the United Nations). 2001. *Global forest resources assessment 2000.* FAO forestry paper 140. Rome: FAO.

Finegan, B., and R. Nasi. 2004. The biodiversity and conservation potential of shifting cultivation landscapes. In *Agroforestry and biodiversity conservation in tropical landscapes.* Edited by G. Schroth, G. A. B. da Fonseca, C. A. Harvey, C. Gascon, H. L. Vasconcelos, and A. N. Izac. Washington, D.C.: Island Press.

Folke, C., L. Pritchard Jr., F. Berkes, J. Colding, and U. Svedin. 1998. *The problem of fit between ecosystems and institutions.* IHDP working paper, no. 2. Bonn, Germany: International Human Dimensions Programme on Global Environmental Change (IHDP).

Geist, H. J., and E. F. Lambin. 2001. *What drives tropical deforestation? A meta-analysis of proximate and underlying causes of deforestation based on subnational case study evidence.* LUCC Report Series, no. 4. Louvain-la-Neuve, Belgium: LUCC International Project Office.

Hecht, S. B. 1985. Environment, development and politics: Capital accumulation and the livestock sector in eastern Amazonia. *World Development* 13(6):663–84.

Klepeis, P. 2003. Development policies and tropical deforestation in the southern Yucatán Peninsula: Centralized and decentralized approaches. *Land Degradation and Development* 14:1–21.

Kummer, D. M. 1992. *Deforestation in the postwar Philippines.* University of Chicago geography research paper, no. 234. Chicago: University of Chicago Press.

Kummer, D. M., and B. L. Turner II. 1994. The human causes of deforestation in Southeast Asia. *BioScience* 44(5):323–28.

Laurance, W. F., M. A. Cochrane, S. Bergen, P. M. Fearnside, P. Delamonica, C. Barber, S. D'Angelo, and T. Fernandes. 2001. The future of the Brazilian Amazon: Development trends and deforestation. *Science* 291:438–39.

Laurance, W. F., P. M. Fearnside, A. K. M. Albernaz, H. L. Vasconcelos, and L. V. Ferreira. 2005. Amazon deforestation models—response. *Science* 307:1044.

Mann, Charles C. 2002. 1491. *Atlantic Monthly*, March, 41–53.

Matthews, E. 2001. *Understanding the Forest Resources Assessment 2000.* Forest Briefing, no. 1. Washington, D.C.: World Resources Institute.

Moran, E. F., and E. Brondizio. 1998. Land-use change after deforestation in Amazonia. In *People and pixels: Linking remote sensing and social science.* Edited by D. Liverman, E. F. Moran, R. R. Rindfuss, and P. C. Stern. National Research Council. Washington, D.C.: National Academy Press, 94–120.

Myers, N. 1992. *The primary source: Tropical forests and our future.* New York: Norton.

National Research Council. 1997. *Rediscovering geography: New relevance for science and society.* Washington, D.C.: National Academy Press.

Perz, S. G., and D. L. Skole. 2003. Secondary forest expansion in the Brazilian Amazon and the refinement of forest transition theory. *Society and Natural Resources* 16:277–94.

Place, S. E., ed. 2001. *Tropical rainforests: Latin American nature and society in transition.* 2nd ed. Wilmington, Del.: Scholarly Resources.

Rappaport, R. 1971. The flow of energy in an agricultural society. *Scientific American* 225:116–32.

Rudel, T. K., D. Bates, and R. Machinguiashi. 2002. A tropical forest transition? Agricultural change, out-migration, and secondary forests in the Ecuadorian Amazon. *Annals of the Association of American Geographers* 92(1):87–102.

Sack, R. D. 1997. *Homo geographicus: A framework for action, awareness, and moral concern.* Baltimore, Md.: Johns Hopkins University Press.

Sadoulet, E., A. de Janvry, and B. Davis. 2001. Cash transfer programs with income multipliers: PROCAMPO in Mexico. *World Development* 29(6):1043–56.

Schaeffer, R., and R. L. V. Rodrigues. 2005. Underlying causes of deforestation. *Science* 307:1046.

Skole, D., W. H. Chomentowski, W. A. Salas, and A. D. Nobre. 1994. Physical and human dimensions of deforestation in Amazonia. *BioScience* 44(5):314–22.

Turner, B. L., II. 1990. The rise and fall of Maya population and agriculture, 1000 B.C. to present: The Malthusian perspective reconsidered. In *Hunger and history: Food shortages, poverty and deprivation.* Edited by L. Newman. Oxford: Basil Blackwell, 178–211.

Turner, B. L., II, J. Geoghegan, and D. R. Foster, eds. 2004. *Integrated land-change science and tropical deforestation in the southern Yucatán: Final frontiers.* Oxford: Oxford University Press.

Turner, B. L., II, P. Klepeis, and L. S. Schneider. 2003. Three millennia in the southern Yucatán peninsular region: Implications for occupancy, use, and "carrying capacity." In *Lowland Maya area: Three millennia at the human-wildland interface.* Edited by A. Gómez-Pompa, M. Allen, S. Fedick, and J. Jimenez-Osornio. New York: Haworth Press, 361–87.

Turner, B. L., II, S. C. Villar, D. Foster, J. Geoghegan, E. Keys, P. Klepeis, D. Lawrence, P. M. Mendoza, S. Manson, Y. Ogneva-Himmelberger, A. B. Plotkin, D. P. Salicrup, R. R. Chowdhury, B. Savitsky, L. Schneider, B. Schmook, and C. Vance. 2001. Deforestation in the southern Yucatán peninsular region: An integrative approach. *Forest Ecology and Management* 154(3):353–70.

Whitmore, T. M., and B. L. Turner II. 2001. *Cultivated landscapes of Middle America on the eve of conquest.* Oxford: Oxford University Press.

Williams, M. 1990. Forests. In *The earth as transformed by human action.* Edited by B. L. Turner II, W. Clark, R. Kates, J. Richards, and W. Meyer. Cambridge: Cambridge University Press, 179–201.

SUGGESTED READINGS

Bray, D. B., and P. Klepeis. Forthcoming. Deforestation, forest transitions, and institutions for sustainability in southeastern Mexico, 1900–2000. *Environment and History.*

Lambin, E. F., and H. J. Geist. 2003. Regional differences in tropical deforestation. *Environment* 45(6):22–36.

Mann, Charles C. 2002. 1491. *Atlantic Monthly*, March, 41–53.

O'Brien, K. L. 1998. *Sacrificing the forest: Environmental and social struggles in Chiapas.* Boulder, Colo.: Westview Press.

Place, Susan E. 2001. *Tropical rainforests: Latin American nature and society in transition.* Jaguar Books on Latin America, no. 2. Wilmington, Del.: Scholarly Resources.

Schmink, M., and C. H. Wood. 1992. *Contested frontiers in Amazonia.* New York: Columbia University Press.

Turner, B. L., II, J. Geoghegan, and D. R. Foster. 2004. *Integrated land-change science and tropical deforestation in the southern Yucatán: Final frontiers.* Oxford: Oxford University Press.

Vandermeer, J., and I. Perfecto. 2005. *Breakfast of biodiversity: The truth about rainforest destruction.* 2nd ed. Oakland, CA: Institute for Food and Development Policy.

HAPTER

5

Tropical Trade-offs

AN ECONOMIC
ERSPECTIVE ON
TROPICAL
FORESTS

Erin O. Sills
and
Subhrendu K. Pattanayak

Introduction

Economics is the study of how to allocate scarce resources in order to maximize benefits. Economists are fundamentally interested in the trade-offs that must be made when there is **scarcity**, that is, when there are limits on resources such as labor or land. Tropical forests, and the land on which they grow, are scarce resources. Allocation of these resources is complicated by the fact that many of the benefits of maintaining forests are not traded in markets, they accrue over the long term, and they are valued by people around the globe. The benefits of clearing forest, on the other hand, are mostly products that can be sold or consumed for the immediate benefit of local landowners. Economists seek to quantify this full set of benefits, understand how people make decisions about forest resources, and suggest policies that will result in higher net benefits.

All economics is essentially anthropocentric, that is, concerned with how humans use resources. Thus, the economic view of tropical forests is as a source of goods (e.g., wood, foods, medicines) and **environmental services** (e.g., soil stabilization, wildlife habitat) that benefit humans. From an economic perspective, **deforestation** is also a process for obtaining valuable goods (e.g., agricultural land) from the forest. With deforestation or any other process, the core issue for economists is whether and how scarce resources are allocated optimally to maximize net benefits to people. **Net benefits** are the value of goods and services (people's **willingness to pay**, in terms of money or some other valuable resource) minus their **opportunity costs** (what people have to pay or what resources they have to invest to obtain the goods and services).

Net benefits can be calculated from two perspectives. First, "private" benefits and costs directly affect families and companies making the decisions about resource allocation. They include goods and services that are bought and sold in the market, as well as those consumed directly. The private net benefits of deforestation are the agricultural returns minus the value of foregone future forest production. Future forest production includes timber that could be sold, as well as goods consumed directly, such as fuelwood. However, these private benefits and costs do not include environmental side effects on others, called "**externalities**." Examples of externalities from deforestation include changes in water flow to farms downstream and loss of habitat for migratory songbirds. The second perspective on net benefits includes these externalities in order to calculate "social," or public, benefits and costs.

When economists study the behavior of individuals, families, or companies (collectively called **economic agents**), they start from the assumption that these agents make rational choices to maximize their private net benefits, given available resources. Choices are also affected by market prices, available **production technology** (methods for combining inputs to produce outputs), government policies and regulations, and information. Microeconomists seek to understand how these conditions affect the decisions of agents, for example decisions about how much land to deforest. Macroeconomists seek to understand the aggregate allocation of resources that results from many agents trading (buying and selling) in markets. The terms of trade are determined by agents' willingness to pay for consumption and their opportunity costs for producing goods and services, as well as institutional rules and policies defined by governments. Collectively, the trades define a set a prices, which in turn provide signals to economic agents about the relative value and scarcity of different goods and services. Because they play such a key role in the functioning of markets and the decisions of agents, prices are of central interest in economics.

Many socially valuable goods and services do not have prices because they are not traded in markets. This is true by definition for externalities: they are external to the market. Economists have defined a category of goods and services called "public"; no one can be excluded from the benefits of **public goods**, and their benefits cannot be divided up and sold. The classic example of a public good is national defense, which benefits everyone in a country. **Biodiversity** and carbon sequestration are examples of public goods provided by tropical forests. Economists believe that the lack of prices for public goods and externalities is a key reason for many environmental problems, including excessive deforestation. Because the agents who decide the fate of the forests are not paid for these social values, they do not include them in their calculation of private net benefits.

In this chapter, we will review economic theories and empirical studies on three aspects of deforestation. First, what factors influence agents' decisions about whether and how much to deforest? Second, what are the consequences of deforestation for society, or in other words, how much are forests worth to the public? Third, what government policies or market incentives will encourage agents to limit deforestation to the socially optimal level?

Theoretical Concepts and Paradigms

Economists often start with a simple conceptual model that represents the essence of a problem. The goal is to find a model that is realistic enough to have explanatory and predictive power but simple enough that it can be tested with available data. Such a model is like a map, containing essential information but certainly not all the details of a photograph. Here, we will use one simple graph to develop theory relevant to the three aspects of deforestation.

As discussed in the introduction to this volume, **deforestation** is the permanent conversion of forest to some other land cover. Thus, an economic theory of deforestation must consider the net benefits of those alternative land covers. In the tropics, forests are most often converted to agricultural land, including annual crops, perennial plantations, and pasture. The economic agents who decide whether, where, and when to convert forest land include agricultural companies, ranchers, communities, and family farmers. When these agents clear forest land, they have some immediate costs (e.g., they must invest time and buy chainsaws) and perhaps some immediate benefits from the timber harvested. They will gain a future stream of net benefits from agriculture and ranching (the value of outputs minus the cost of inputs). To compare all of the costs and benefits, the future values are adjusted by the **discount rate** to find their equivalent present value. The sum of discounted benefits minus costs on a per-acre basis is the "**marginal net benefits of agriculture**" (MNB_A) shown in figure 5.1. This represents the dollar value of one additional hectare of cleared land, or the value of the "marginal" hectare. The total value of agriculture is the area under the curve up to a chosen level of deforestation (D_P*), shown on the horizontal axis.

To understand why MNB_A declines as deforestation increases, consider the case of a family farmer who owns 100 hectares of forest. In this case, the origin (left side) of the graph represents 100 hectares of forest, and the right side represents 100 hectares of cleared land. The benefits of clearing and planting the first hectare in crops are very high, because this provides

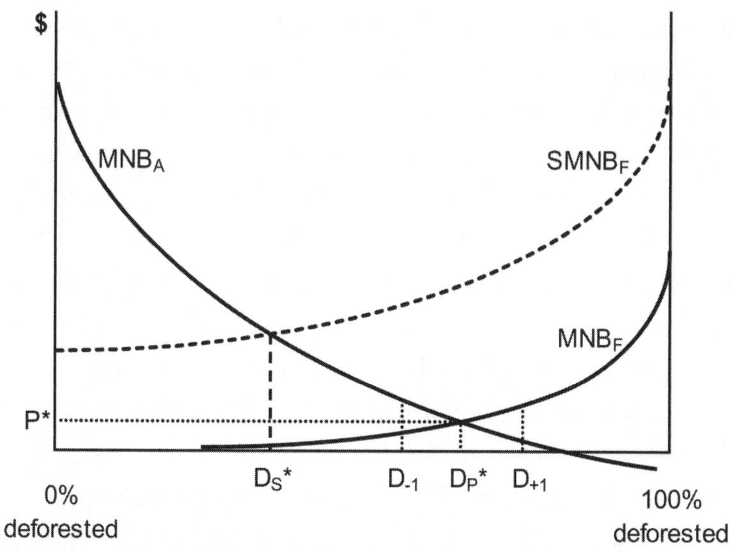

Figure 5.1. Marginal Net Benefits of Agriculture

necessary food for the family. As more land is cleared, the MNB_A falls, partly because the value of crops falls (for example, once the family has sufficient food and must seek a market for additional production) and partly because costs rise (for example, if the family must take children out of school or start hiring workers). Eventually, the costs become greater than the benefits, and the MNB_A falls below zero.

Now consider the "**marginal net benefits of forest**," or MNB_F, curve. When the farmer has 100 hectares of forest, it is not scarce, and consequently its marginal value is zero: the farmer does not gain anything from having an extra hectare of forest. However, as forest becomes scarce, its value increases. When nearly all of the land is cleared, one hectare of forest provides significant benefits in terms of future production of fuelwood, vines, fruits, game, and the like.

The farmer clears forest as long as MNB_A is greater than MNB_F, because this brings net benefits. The farmer stops deforesting when MNB_F rises above MNB_A, and thus D_P* represents the total number of hectares cleared. This level of deforestation is optimal for the farmer, or **efficient**, because she could not do any better by choosing a different level of deforestation.

By clearing fewer hectares (D_{-1}), the farmer would give up gains in crop production that are greater than the loss in forest production. By clearing more hectares (D_{+1}), the farmer would give up forest benefits greater than the gains in crop production. D_P* maximizes the total net benefits, which is the area under the MNB_A from the left to D_P* plus the area under MNB_F from D_P* to the right.

We can also reinterpret the graph as showing the percent of a region that is deforested. Consider a region where land can be bought and sold. Forest owners sell their land if they can get a price per hectare higher than MNB_F, and farmers buy land if the price per hectare is less than MNB_A. Thus, landowners trade until MNB_F and MNB_A are just equal, again resulting in D_P* hectares cleared for farming. At that point, the price of land will be P*, just equal to the marginal value of land in agriculture and in forest. Again, this maximizes benefits, because all land is allocated to the use that yields the highest net benefits. To see this, pick a level of deforestation other than D_P* and calculate the changes in forest and agricultural benefits (the areas under the MNB_F and MNB_A curves); the loss will always exceed the gain.

This theory suggests that in order to understand deforestation—whether by individuals or in an entire region—we must identify the factors that determine the MNB_F and MNB_A curves. There are several different approaches to this problem. The **Ricardian approach** focuses on the inherent profitability of the land, as characterized by physiographic properties such as soil quality, elevation, and precipitation. In the **von Thünen approach**, distance to markets is considered the key determinant of land use. This suggests the importance of roads constructed by governments and logging companies. A third approach focuses on the farm family as both producer and consumer of forest products and crops. This **household production approach** recognizes that one determinant of value is family demand for outputs. This demand in turn depends on socioeconomic characteristics such as family size, income, wealth, and cultural background. Combining these different approaches, a complete economic model of deforestation would include biophysical characteristics of the land and forest, location, socioeconomic characteristics of the landholder, prices, and available technology, which in turn may be influenced by government policies and regulation.

These are the direct factors that a family farmer or other landholder considers when deciding how much to deforest. In turn, these factors are set by underlying conditions such as extent of the road network, exchange rates and trade policy, research and development, population density, and conditions in other regions where people are deciding whether to migrate to the forest frontier. At an aggregate level, the extent of deforestation depends on these underlying conditions that define the decision-making environment for individual agents.

Thus far, we have discussed economic models for explaining and predicting deforestation. We have shown how individual landholders and land markets reach levels of deforestation that maximize private net benefits. However, other chapters in this volume argue that tropical deforestation is not **socially optimal** and that the loss of forest goods and services far outweighs the gains from alternative land uses. We can use figure 5.1 to understand this difference in perspectives. Recall that MNB_A and MNB_F represent *private* net benefits. If the social benefits of forests are higher or the social benefits of agriculture are lower, then the optimal level of deforestation from a social perspective will be lower than D_P^*. The dashed curve in figure 1 shows higher **social marginal net benefits** of forest ($SMNB_F$). Try drawing a lower $SMNB_A$ curve and recalculate the optimal level of deforestation, D_S^*.

Consider first why social net benefits from forest are likely to be higher than the private net benefits. One key reason is that many outputs of tropical forests are public goods. The loss of some of these goods, such as extinction of species, is irreversible, and as a result the public value of preserving these goods is very high. Private landholders cannot capture these values and hence do not take them into account in decisions about deforestation. Another reason for differences in social and private MNB_F is that private landowners may have limited information and training in forestry. For example, colonists who settle tropical forest frontiers are often displaced farmers from other regions, and they are well versed in crop production but not in forest management. Thus, they may not know how to capture all the benefits of forests.

A third set of reasons is related to differences in private and social valuation of the future flow of goods and services from forests. From a social perspective, forest outputs a hundred or even a thousand years from now may be quite important, but for individual landholders, these distant benefits are often outweighed by current concerns. This may be because discount rates are higher for individual agents than for society as a whole. It may also be due to lack of access to credit on forest frontiers; if landholders cannot obtain loans, they may not be able to survive in the short term while they wait for long-term returns from the forest. Finally, in many tropical forest regions, individuals may not be able to obtain secure ownership of forested land. Deforestation may be the only way to secure de facto **property rights** and in some cases may be the best way to obtain formal title to land. In these circumstances, private landholders will consider forest to have zero value, because they cannot secure future forest outputs. All of these factors would lead individual agents to ignore the future benefits of forests, resulting in a lower private MNB_F.

The marginal value of agriculture may also be different from social and private perspectives. The private MNB_A is increased when governments support agriculture with public tax dollars by subsidizing fertilizer, building road networks, giving tax breaks for new agricultural enterprises, and granting title to public land when it is cleared for agriculture. These are called "**perverse incentives**" because they add to the private profitability of agriculture, but from a social perspective, they are simply transfers from taxpayers to farmers and ranchers.

To find the socially optimal level of deforestation, we must quantify the social values of forest and agriculture. For MNB_A, this requires netting out public subsidies to find the social net benefits. For MNB_F, the task is more complex, requiring quantification of the value of public goods and services over long time horizons. These public goods are not traded in markets and so do not have prices. Economists have developed **nonmarket**

valuation methods to discover how much people would pay if they had the opportunity to buy these goods. The total willingness to pay for public goods can then be added to private forest values to find the net social benefits of forest.

Policies are also influenced by the distribution of benefits from forest and agriculture. The benefits of agriculture usually accrue to a small number of farmers and ranchers. On the other hand, the benefits of forests accrue primarily to two groups: global citizens who, in aggregate, place a high value on forest services such as biodiversity, and forest-dependent indigenous people who live in the forest but often do not have title to the land.

Policies to address deforestation must take these distributional issues into account. From an economic perspective, the key problem is that agents of deforestation make their decisions based on the private—not the social—values of forest and agriculture. In other words, farmers, ranchers, and companies operating in tropical forest regions face the "wrong" incentives. This is common to many environmental problems, and environmental economists have identified two categories of policy response: regulatory and market based. If a government could identify the optimal D_S*, it could make rules to fix the maximum level of deforestation. For example, the government could create parks to protect the rest of the forest, where $SMNB_F$ is greater than $SMNB_A$. There are at least two difficulties with this approach, one practical and one ethical. The practical problem is the difficulty of enforcing regulations, such as protected area boundaries, in remote forest frontiers. The ethical problem is the unequal distribution of benefits and costs. In particular, the costs of restricting deforestation are likely to be primarily regional—affecting farmers and ranchers—while the benefits are spread across the entire globe.

Market-based policies are designed to address these problems. We showed above that agents operating in a market will trade until they end up at the optimal allocation as defined by the marginal benefit curves that they face (D_P* in figure 5.1). Market-based policies seek to change those marginal benefit curves by compensating landholders for public forest benefits or charging them for the public costs of agriculture. For example, governments could apply a higher tax rate to deforested lands. This approach has also been adopted by **nongovernmental organizations** (NGOs). For example, there are several initiatives to encourage consumers to pay more for tropical timber that is produced on a sustainable basis from well-managed forest. We will expand on this discussion of policy options in the conclusion to this chapter.

Current Research and Major Findings

DETERMINANTS OF DEFORESTATION

In this section, we review empirical economic research that tests and quantifies the influence of direct factors and underlying conditions on deforestation. The economic theory reviewed in the previous section suggests that agents—such as family farmers—respond to factors that directly determine their private net benefits from forest and alternative land uses. Underlying these factors are **macroeconomic** conditions, demographic trends, and government policy. At a regional or national level, patterns of deforestation can be related to these underlying conditions. It is important to distinguish these direct factors and underlying conditions from the alternative land uses that replace forests. Agents may deforest in order to obtain land for annual or perennial crops, plantations, or pasture. In forests with high densities of valuable timber, agents may harvest enough timber to deforest an area. In dry forests near large population centers, agents may deforest through collection of fuelwood. Identifying the share of forest being converted into these different land uses is an important research question. Economists, however, focus on *why* forests are converted to these different land uses.

To address this question, economists have applied two general methodologies at three different scales. The first methodology is **regression**. This requires large data sets with at least one measure of deforestation and several measures of factors or conditions. The researcher estimates a regression model to test whether the factors or conditions are statistically related to deforestation in that data set. The second methodology is simulation. In this case, the researcher constructs a model of a system using the best available estimates of relationships in the system. For example, if the system were a farm, the relationships would include the effect of price changes on crop selection. Estimates of these relationships are often taken from regression models, and hence the two methodologies are complementary. Both can be used to model deforestation at a national scale, at a regional (county or province) scale, or at an agent (household) scale.

Determinants at a National Scale

Regression models of deforestation levels and rates across different countries constitute what Barbier and Burgess (2001) call the "first wave" of economic studies of deforestation. Based on a review of these studies, they conclude that the following conditions are correlated with deforestation: income; population growth/density; agricultural prices/returns; agricultural yields; agricultural exports/export share; logging prices/returns; roads and road building; scale factors (size of forest stock, land area, etc.); and institutional factors (political stability, property rights, rule of law, etc.) (Barbier and Burgess 2001, 417). However, these conditions are not always correlated with deforestation in the same direction: one researcher might find that a given condition increases deforestation, while another finds that it decreases deforestation. This may be due to differences across regions or time periods or to **econometric** issues.

Consider population as an example. Areas with high population densities generally have experienced more deforestation. However, this might be because some other condition, like agricultural yields, has encouraged both population growth and deforestation. In econometrics, this is called the "omitted variable" problem. A second possible econometric issue is "endogeneity." Rather than population driving deforestation, deforestation may encourage immigration to a region, hence increasing population.

A third issue is related to the data source. Most cross-country regressions rely on data from the Food and Agriculture Organization (FAO) of the United Nations. Historically, FAO has compiled statistics on forest cover reported by country governments. When these statistics were not available for a particular time period, FAO predicted forest cover with a model that included population. Hence, population is related to deforestation in these data sets simply because of the way they were constructed.

The influence of income (usually measured as gross domestic product per capita) on deforestation is of particular interest to economists. Income could be positively associated with deforestation because higher income allows purchase of deforestation inputs such as tractors and chainsaws, and because it expands demand for agricultural products. Income could also be negatively associated with deforestation because higher income allows investment in more intensive agriculture and in the industrial and service sectors, and because it increases demand for the environmental services and products provided by standing forests (see Foster and Rosenzweig 2003). In many developed countries, environmental damage—such as deforestation—did initially increase and then decrease as income grew over time. In fact, forest area has been increasing in countries such as the United States. This pattern is called the "**environmental Kuznet's curve**" (Grossman and Kruger 1995). As with population, evidence for this pattern is mixed. The turning point, where deforestation rates start to fall, appears to vary across countries and to depend on other conditions, such as the distribution of wealth and political freedoms (Bhattarai and Hammig 2001).

Many tropical countries have large international debts due to loans taken out by their governments. The empirical evidence on the relationship between debt and deforestation is mixed. It is clear, however, that the terms of trade facing a country are important to deforestation rates. Currency exchange rates are one determinant of how competitive a country's exports are in the world market. When a country's currency is devalued, its exports become less expensive for other nations. Thus, currency devaluation is often recommended for indebted countries that need to increase export revenue. Currency devaluation is also associated with increased deforestation. This may be a direct result of increased demand for agricultural exports. It may also be the result of increased demand for timber exports, which leads timber companies to expand logging roads further into the forest, improving access for other agents. Simulation models, such as computable general equilibrium models, generally confirm these effects, showing that devaluation, trade liberalization, and lower agricultural export taxes increase deforestation (Kaimowitz and Angelsen 1998, 64).

Determinants at a Regional Scale

The first wave of economic research focused on cross-country studies largely because of the availability of national data. Recently, more economists are studying deforestation across regions within a country by combining census data with increasingly accessible remote sensing data (Pfaff 1999). As with cross-country studies, some researchers have found that population density and income per capita have positive effects on deforestation, while others have found no discernible effect once they control for other conditions. Where data are available, higher agricultural prices are generally found to be associated with more deforestation. For example, Angelsen et al. (1999) find that deforestation is consistently positively related with several different indices of agricultural output prices in Tanzania. They argue that this is evidence that farmers are seeking to maximize their net benefits and not simply reach some minimum level of income.

At the regional scale, models are often based on Ricardo's and von Thünen's approaches, relating deforestation to physiographic factors (soil quality, slope, precipitation) and access (distance to roads and markets). Access is also a proxy for price: better access means higher output prices and lower input prices. This limited set of factors has been found to explain the probability of deforestation at a given point with a high degree of accuracy. In particular, proximity to roads and to previously deforested areas is highly correlated with new deforestation. Chomitz and Thomas (2003) find that distance to roads and precipitation are negatively correlated with deforestation in the

Brazilian Amazon. In Thailand, Cropper et al. (2001) find that physiographic factors have the strongest correlation with deforestation, but roads and population density are also positively related to deforestation.

In addition to regression analysis, economists have developed simulation and analytical models of deforestation at the regional level. One topic that has attracted particular interest is the role of timber prices and exports. Many tropical countries have outlawed the export of whole (unprocessed) logs, partly in an effort to reduce pressure on forests. This generally decreases timber prices, with several potential contradictory effects on deforestation. Referring to figure 5.1, log export bans could shift down MNB_A because of higher access costs due to fewer logging roads and because of lower prices for any timber sold when the land is cleared. On the other hand, the MNB_F would also fall because of lower timber prices. While early economic studies concluded that log export bans were generally bad policy because they reduced MNB_F, more recent studies have concluded that the effect on MNB_F is small where tenure is insecure and discount rates are high (von Amsberg 1998). In these cases, log export bans might slow deforestation by reducing MNB_A.

Determinants at the Agent Scale

Perhaps the most intuitive approach to modeling deforestation is to examine the behavior of the agents who actually decide whether to deforest an area (Caviglia-Harris and Sills, forthcoming). Regression analysis at this scale generally requires primary data from surveys of farm households. These surveys usually include questions about the extent of deforestation (number of hectares or percent of landholding) as well as about the factors hypothesized to influence deforestation. Current research focuses on linking such survey data with measures of deforestation from remote sensing (Fox et al. 2002).

Regression results generally show that improved access to market centers and to roads leads to increased deforestation. Prices of agricultural outputs are also positively associated with deforestation (Coxhead et al.

2001). In many countries, agricultural prices fluctuate substantially, and there is some evidence that small farmers increase deforestation in response to this uncertainty, seeking to reduce risk by producing their own food or by investing in perceived safe income sources like cattle, rubber, and coffee. Among input prices, wages (the price of labor) have been found to be the most influential. Higher wages generally decrease deforestation, both because of the increased cost of hiring additional workers and because high wages attract family workers into the labor market (Shively 2001).

Where there is no labor market, family size is a key factor positively correlated with deforestation. A number of studies have focused on demography (family life cycle) as a key determinant of land use decisions (Walker et al. 2002). During a family's life cycle, the balance between labor (available workers) and consumers (children and elderly) evolves, with consequences for risk aversion, discount rates, and market integration. Walker and colleagues (2002) summarize regression results for life cycle and other socioeconomic factors from studies in the Amazon. For example, income has been found to be positively associated with deforestation, although this could be endogenous (a *result* as much as a *cause* of deforestation).

Another theme of recent research is the effect of technology on deforestation rates. A common policy prescription for reducing deforestation is development and dissemination of technologies that will allow farmers to increase production on land that is already deforested. Economists, however, raise the concern that this will simply make it even more profitable to deforest (by raising MNB_A). For example, Vosti et al. (2002) use a simulation model of farms in the western Brazilian Amazon to show that improved ranching and farming technology would increase the rate of deforestation. Deforestation might be reduced by some forms of improved technology, such as more profitable techniques for sustainable forest management, infrastructure like irrigation that is only viable in nonforested regions (Shively 2001), and possibly labor-intensive agricultural techniques. In general, technologies that use more scarce resources—like labor—will constrain deforestation in the short run, although in the long run they may simply attract more of the scarce resource to the region (e.g., through migration).

In sum, economists have found that the determinants of deforestation are complex and location specific. The most consistent result is that deforestation increases when agricultural profits increase, whether that is due to improved access (usually due to expanded road networks) or higher prices for outputs (in domestic or export markets). To evaluate specific policy interventions in specific regions, studies applying different methodologies at all three scales are useful. For example, a national simulation model could predict how prices and wages would adjust to a policy change; a regression model of agents could predict how they would respond to these new prices; and a regional model could predict resulting changes in the spatial pattern of deforestation.

CONSEQUENCES OF DEFORESTATION

The consequences of deforestation for decision makers—that is, the private costs and benefits—were already taken into account in the previous models of agent behavior. Most of the benefits of deforestation are private. Income from agriculture, including timber sales from areas being cleared, is generally earned by the deforestation agents. However, many of the costs of deforestation are externalities—they fall on other people. These costs are basically the loss of forest goods and services, or $SMNB_F$. Here, we consider these costs in three categories: local, global, and national.

Local Consequences

Throughout the tropics, forests are legally owned by governments but are used by indigenous and other traditional rural populations. Because these people do not own the forest, the benefits they derive from the forest are often not considered in deforestation decisions. Although there are no precise estimates, researchers agree that hundreds of millions of rural people rely on tropical forests for a variety of products and services (Byron and

Arnold 1999). In case studies around the world, economists have quantified the value of nontimber forest products (NTFPs) used as food, medicine, fuel, construction materials, and agricultural inputs such as fodder for livestock. In many cases, the value of these products—estimated at market prices or in terms of willingness to trade with market products—is quite low per hectare of forest compared to the value of timber and agricultural use of the land. However, the value of NTFPs as a percent of household income can be significant, especially for the poorest households.

For some indigenous peoples, the forest is central not only to their livelihoods but also to their cultural and spiritual well-being. At the other end of the spectrum, many small farmers obtain only a small fraction of their income from the forest, but they rely on it as an important safety net in case of drought or other crisis (Pattanayak and Sills 2001). In much of Africa, forest products are known as "famine foods," which people rely on when their crops fail. The spiritual and safety-net roles of forests may explain why traditional peoples fight to protect and maintain access to them, even when they derive most of their income from other sources.

There are also important instances of local people earning the majority of their cash income from forests by collecting and selling NTFPs to international or urban markets. Examples include Brazil nuts in the Amazon, rattan (a climbing palm) in Southeast Asia, and bush meat (game meat) in west Africa.

In addition to products, forests provide **ecosystem services**—conditions and processes that sustain and fulfill human lives (Daily 1997). In many tropical forest regions, loss of these services through deforestation is highly detrimental to local livelihoods. Deforestation may destabilize hydrological systems, leading to flooding and droughts, and contribute to erosion, thus clogging irrigation systems and damaging fisheries. While forests everywhere play an important role in controlling erosion, it is difficult to generalize about other **watershed** services, because the impact of forests on hydrological systems varies according to climate, geology, and forest type. One study in Indonesia quantified the value of forests to downstream farmers by comparing agricultural profits across watersheds with different levels of water flow associated with different amounts of deforestation, controlling for other factors that affect profits (Pattanayak and Kramer 2001). In this case, forests make valuable contributions to local livelihoods through watershed protection. Another local forest service that is often cited but rarely quantified is protection from disasters, such as landslides in the wake of hurricanes in Central America and flooding from cyclones in south Asia.

Global Consequences

Many forest ecosystem services are of global value, with the two most prominent being carbon sequestration and biodiversity. Tropical forests and their soils hold an enormous stock of carbon, which would contribute to climate change if released through deforestation. Hence, the value of this service is equivalent to the cost of the avoided climate change (see volume on climate change in this series). Tropical forests have the greatest species diversity of any terrestrial **ecosystem**. Some people value the opportunity to experience this biodiversity through nature tourism and recreation. People also value biodiversity as a potential source of new drugs, genetic material for crops, and other products. In addition, biodiversity has inherent value for some people, who derive pleasure from its very existence. Economists recognize this "existence value" as one type of **nonuse value**. Other nonuse values include "bequest value," which is the value of passing something on as an inheritance to future generations, and "option value," which is the value of keeping open the option to use something in the future. Donations to "save the rainforest" campaigns by people who have never visited and do not intend to visit the tropics are evidence for these nonuse values.

Because they are not related to any behavior, the only way to quantify these nonuse values is by directly asking people in a survey. The most common survey method is "contingent valuation," in which a hypothetical payment mechanism is posed, and respondents are asked how much they would be willing to pay. For example, Kramer and Mercer (1997) asked a random sample of U.S. residents about their willingness to pay a donation to a hypothetical United Nations fund to preserve rainforests. Horton et al. (2003) asked residents of Italy and the UK about their willingness to pay additional EU (European Union) taxes to support a new protected area system in Brazil. Both of these studies found aggregate values large enough to greatly expand and maintain protected areas throughout the tropics. Some economists are critical of the contingent valuation method because it relies on people to accurately and honestly report their willingness to pay, rather than on observing their behavior (i.e., what prices they pay for goods in the market). When interpreting the results of a contingent valuation study, it is important to critically evaluate both the question posed in the survey and the sample of people interviewed.

National Consequences

At the national level, the costs of deforestation are a combination of the aggregated local costs; the appropriate fraction of global costs; and the lost ecosystem services that specifically affect the nation, such as silting of reservoirs, smoke from forest fires, and degradation of landscapes that are key tourist attractions. Economists have employed a variety of nonmarket valuation methods to quantify these services. For example, with the "dose-response" method, regression models are used to link smoke to respiratory illness, and with the "travel cost" method, tourism values are deduced from the distance that tourists are willing to travel to visit the tropical forest. Another approach to national values focuses on adjusting national accounts, such as gross domestic product, to reflect changes in natural resource stocks and environmental quality. For example, in one of the earliest applications of this so-called green accounting, Repetto et al. (1989) calculated that Indonesia's growth rate in the 1970s and 1980s was only 4%—rather than 7%—once depletion of forest and soil resources had been taken into account. This raises the question of why government policy often appears to ignore these very significant nonmarket forest values.

When economists sum up costs and benefits across people to find net benefits, they often do not consider the distribution of those costs and benefits. The assumption is that if the total net benefit is positive, then those who "win" can compensate those who "lose." For example, if conservation of all remaining forests in a country has positive net benefits summing across all gains and losses, this would be judged an appropriate policy. The practical difficulty is implementing a policy in which the "winners" compensate the "losers." The many winners from forest conservation are spread across the globe, while the losers are the relatively few agents of deforestation (van Beukering 2003). Thus, compensation schemes would likely have high **transactions costs**, because payment must be collected from a large, diffuse group of beneficiaries. Further, the groups that have the most to gain from deforestation often have the most political influence and are willing to invest the most in lobbying. This encourages government officials to allow deforestation even if there are net costs to society as a whole.

Policy Options and Conclusions

A key insight of the economic perspective on deforestation is that some level of deforestation is socially optimal. Unless that level has already been exceeded in a region, the goal of policy should be to limit—but not stop—deforestation. The challenge is that the optimal level varies across scales. Many of the costs of deforestation are shared by all of the world's citizens, while the benefits accrue to a relatively small group of agents. This suggests that some compensation scheme will be necessary for progress toward the globally optimal level of deforestation.

Empirical economics research consistently finds that deforestation is driven partly by the two interrelated factors of roads and prices. Road-building projects merit the attention they receive from concerned environmental groups. On the other hand, there are good reasons for governments to build roads: they increase the private benefits to agriculture and facilitate health and education services, thus enhancing the income and well-being of many relatively poor constituents, such as family farmers on the forest frontier. These farmers respond to the new input and output prices induced by roads to maximize their net benefits. The problem is that from a public perspective, these are the "wrong" prices. To get the prices right, some mechanism is required to translate the public benefits of forests into private market incentives.

One remedy for the market's failure to reflect the public value of forests is to turn forest goods and services into tradable commodities, such as carbon sequestration credits, bioprospecting contracts, and conservation concessions (Landell-Mills and Porras 2002). Trade in these new commodities requires that governments establish appropriate regulatory frameworks that reduce transactions costs—for example, clear and enforceable property rights to the commodities and to forestland. This does not depend solely on governments. Environmental organizations can act as brokers in these markets, matching up enough buyers and sellers to reach the minimum threshold of forestland required to provide an ecosystem service such as species conservation or watershed protection. The distributional implications of this market-driven approach will depend largely on how the initial property rights are distributed: who has the right to sell the carbon, biodiversity, and other conservation values of the forest?

Nongovernmental organizations are also involved in a second market-driven approach to deforestation: bundling forest services with existing commodities through systems of audits, certification, and labeling. For example, ecotourism, timber, and NTFPs such as Brazil nuts are all commodities with established markets, and they can all be produced under forest management systems that sustain other valued forest services. Landowners could profit from forest services by charging a premium for forest commodities labeled as having been produced through sustainable forest management. In practice, many consumers are not willing to pay a large price premium, although they will choose labeled products if they are in the same price range. Thus, the benefits of labeling to producers may be expansion of market share and avoidance of negative publicity, rather than an actual price premium.

While these market-driven approaches focus on raising the private benefits of forests, the other half of the equation—the private benefits of agriculture—is an equally important policy concern. The excess of private over public benefits from agriculture is usually due to government policy, including **subsidies**, tax breaks, and other perverse incentives, such as awarding tenure as a function of deforestation. Taxing cleared land at a higher rate than forestland would be one way to address this. However, policies and economic conditions outside of the forest frontier—including exchange rates, energy policy, and land distribution—are often key determinants of the factors that enter into decisions by deforestation agents. For example, wage rates are determined by macroeconomic conditions and economic policies that have no apparent relation to forestry. Increasing wage rates, however, is one effective way to raise the opportunity cost—and hence lower the net benefits—of agriculture and deforestation (Shively 2001).

Economists give mixed verdicts on three other policies that are widely advocated. First, consider bans on exporting unprocessed logs of tropical timber (or boycotts of log imports). As discussed above, many economists have criticized this approach because it lowers the private benefits of forest. However, others have pointed out that this approach may be an effective way to slow expansion of the road network and hence raise the cost of clearing land for agriculture. Second, consider programs to provide credit and secure land tenure for long-term forest management. Again, economists predict two conflicting effects. Secure tenure and credit could make it feasible to invest in forest management, but it could also raise the returns to investments in deforestation and agriculture. The outcome will depend on relative returns, which in turn depend on biophysical factors (e.g., soil quality) and expertise of landowners (e.g., ranchers versus traditional forest people). Third, consider agricultural intensification through improved technology. Economists point out that the net effect depends on the broader economic context: as long as there is a large potential market for agricultural products, raising agricultural productivity in the forested region is likely to encourage deforestation. The net effect also depends on the specific technology: if it is only applicable to nonforested regions or if it uses up a scarce input like labor, then it is more likely to act as a brake on deforestation.

Better agricultural technology, higher wage rates, and increased market demand for ecosystem services are all characteristics of developed countries. This suggests an alternative perspective on deforestation policy based on the environmental Kuznet's curve. Perhaps to reduce deforestation, we should promote development. Great care must be taken with this argument, because losses of some forest goods and services cannot be reversed at later stages of development. Further, the environmental Kuznet's curve pattern is a function of the political, social, and institutional aspects of development as well as pure economic growth. Thus, we cannot rely on economic growth alone. Protecting forest in parks and by government regulation of land use should also be part of the policy package to address deforestation. Protected areas may even promote the development of markets by "closing the forest frontier" and forcing agents to confront the fact that forest resources are scarce. The economic perspective on deforestation policy suggests combining this direct approach of protecting forests in parks with both the long-term approach of promoting economic development and the market approach of providing the right incentives to the economic agents who make deforestation decisions.

Acknowledgments

Research assistance was provided by Saeko Makino.

REFERENCES

Angelsen, A., E. F. K. Shitindi, and J. Aarrestad. 1999. Why do farmers expand their land into forests? Theories and evidence from Tanzania. *Environment and Development Economics* 4:313–31.

Barbier, E., and J. C. Burgess. 2001. The economics of tropical deforestation. *Journal of Economic Surveys* 15(3):413–33.

Bhattarai, M., and M. Hammig. 2001. Institutions and the environmental Kuznetus curve for deforestation: A crosscountry analysis for Latin America, Africa and Asia. *World Development* 29(6):995–1010.

Byron, N., and M. Arnold. 1999. What futures for the people of the tropical forests? *World Development* 27(5):789–805.

Caviglia-Harris, J., and E. Sills. Forthcoming. Land use and income diversification: Comparing traditional and colonist populations in the Brazilian Amazon. *Agricultural Economics.*

Chomitz, K. M., and T. S. Thomas. 2003. Determinants of land use in Amazônia: a fine-scale spatial analysis. *American Journal of Agricultural Economics* 85(4):1016–28.

Coxhead, I., A. Rola, and K. Kim. 2001. How do national markets and price policies affect land use at the forest margin? Evidence from the Philippines. *Land Economics* 77:250–67.

Cropper, M., J. Puri, and C. Griffiths. 2001. Predicting the location of deforestation: The role of roads and protected areas in North Thailand. *Land Economics* 77(2):172–86.

Daily, G. 1997. What are ecosystem services? In *Nature's services: Societal dependence on natural ecosystems.* Edited by G. Daily. Washington, D.C.: Island Press, 1–10.

Foster, A. D., and M. D. Rosenzweig. 2003. Economic growth and the rise of forests. *Quarterly Journal of Economics* 118:601–37.

Fox, J., R. R. Rindfuss, S. J. Walsh, and V. Mishra, eds. 2003. *People and the environment: Approaches for linking household and community surveys to remote sensing and GIS.* Dordrecht, the Netherlands: Kluwer Academic Publishing Group.

Grossman, G., and A. Kruger. 1995. Economic growth and the environment. *Quarterly Journal of Economics* 110(2):353–77.

Horton, B., G. Colarullo, I. J. Bateman, and C. A. Peres. 2003. Evaluation of non-user willingness to pay for a large-scale conservation programme in Amazonia: a UK/Italian contingent valuation study. *Environmental Conservation* 30(2):139–46.

Kaimowitz, D., and A. Angelsen. 1998. Economic models of deforestation: A review. Bogor, Indonesia: Center for International Forestry Research.

Kramer, R., and E. Mercer. 1997. Valuing a global environmental good: U.S. residents' willingness to pay to protect tropical rain forests. *Land Economics* 73:196–210.

Landell-Mills, N., and I. T. Porras. 2002. Silver bullet and fool's gold? A global review of markets for forest environmental services and their impacts on the poor. London: International Institute for Environment and Development.

Pattanayak, S. K., and R. Kramer. 2001. Worth of watersheds: A producer surplus approach for valuing drought control in Eastern Indonesia. *Environment and Development Economics* 6:123–45.

Pattanayak, S. K., and E. Sills. 2001. Do tropical forests provide natural insurance? Non-timber forest product collection in the Brazilian Amazon. *Land Economics* 77:595–612.

Pfaff, A. 1999. What drives deforestation in the Brazilian Amazon? *Journal of Environmental Economics and Management* 37(1):26–43.

Repetto, R., et al. 1989. *Wasting assets: Natural resources in the national income accounts.* Washington, D.C.: World Resources Institute.

Shively, G. 2001. Agricultural change, rural labor markets, and forest clearing: An illustrative case from the Philippines. *Land Economics* 77:268–85.

van Beukering, P. J. H., H. S. J. Cesar, and M. A. Janssen. 2003. Economic valuation of the Leuser National Park on Sumatra, Indonesia. *Ecological Economics* 44:43–62.

von Amsberg, J. 1998. Economic parameters of deforestation. *World Bank Economic Review* 12:133–53.

Vosti, S., J. Witcover, and C. Carpentier. 2002. Agricultural intensification by smallholders in the western Brazilian Amazon. Research report 130. Washington, D.C.: International Food Policy Research Institute.

Walker, R., S. Perz, M. Caldas, and L. T. Silva. 2002. Land use and land cover change in forest frontiers: The role of household life cycles. *International Regional Science Review* 25:169–99.

SUGGESTED READINGS

Andersen, L. E., C. W. J. Granger, E. J. Reis, D. Weinhold, and S. Wunder. 2002. *The dynamics of deforestation and economic growth in the Brazilian Amazon.* New York: Cambridge University Press.

Angelsen, A., and D. Kaimowitz. 1999. Rethinking the causes of deforestation: Lessons from economic models. *World Bank Research Observer* 14(1):73–98.

Heal, G. 2000. *Nature and the marketplace: Capturing the value of ecosystem services.* Washington, D.C.: Island Press.

Sandler, T. 1993. Tropical deforestation: Markets and market failures. *Land Economics* 69(3):225–33.

Sills, E., S. Lele, T. Holmes, and S. Pattanayak. 2003. Role of nontimber forest products in the rural household economy. In *Forests in a Market Economy*. Edited by E. Sills and K. Abt. Dordrecht, the Netherlands: Kluwer Academic Publishing Group, 259–81.

Southgate, D. 1998. Tropical forest conservation: An economic assessment of the alternatives in Latin America. New York: Oxford University Press.

Wibowo, D. H., and R. N. Byron. 1999. Deforestation mechanisms: A survey. *International Journal of Social Economics* 26(1–3):455–74.

HAPTER

6

Global Governance

INTERNATIONAL RELATIONS ERSPECTIVE ON TROPICAL FORESTS

Doris Fuchs

Introduction

Harold Lasswell once described the study of politics as the study of "who gets what, when, and how" (Lasswell 1958). Others, such as David Easton, have described political science as the inquiry into the structures and processes that determine "the authoritative allocation of values for a society" (Easton 1971). Although these definitions differ slightly, they provide a general characterization of the study of politics, which requires investigation into the causes and consequences of political, economic, and social activity.

Defining politics as a struggle for control among competing interests, political scientists traditionally have emphasized "power" as the key concept

(Lasswell and Kaplan 1950). In the last decades, this focus was consciously expanded to take better into account the (complementary or conflictual) interests of actors, as well as the influence of ideas. From this political perspective, tropical **deforestation** can be understood as the influence and interaction between regulatory frameworks shaped by powerful actors, their economic decisions, and the popularity of certain images and ideas associated with tropical forests.

In recent years, the fate of the world's tropical forests has captivated the media and people around the world. This public attention has increased the pressure for states—even those that have no tropical forest resources—to find international means to save these unique **ecosystems**. As with other global environmental problems, international efforts to save the world's tropical forests have been complicated by issues of state sovereignty, the complexity of domestic and international economic forces, and the varied objectives of powerful international actors. In this chapter, we review international efforts to address tropical deforestation and the promise of public–private partnerships for curbing rates of this deforestation.

Development of an International Forest Regime

Political scientists have increasingly explored developments at the international level and their impact on deforestation, examining how **transnational** and **supranational actors**, global markets, and **international norms** affect deforestation patterns as well as national norms and policies. Of interest is the global mobility of capital and its potential impact on rates of resource extraction as well as on domestic environmental policies and standards (see Sills and Pattanayak, this volume). For instance, supranational actors such as the **World Bank** and the **International Monetary Fund** (IMF) have visibly started to intervene in national forestry policy by including specific policy elements in the **conditionality packages** associated with lending. Researchers also note that national governments must increasingly justify changes in domestic economic, social, and environmental policies in light of changes in the global norms of **human rights** and **social justice**—norms consistently at issue in tropical forests, where many indigenous people remain dependent on tropical forest resources. With heightened global concern regarding the fate of tropical forests and the increasing international variables affecting forest management, political scientists have expended considerable efforts in trying to assess and explain the lack of success in negotiations for a global forest **regime**.

An international regime refers to a combination of accepted norms and principles, formal and informal agreements, and the institutions developed to address a specific transnational problem of concern to more than one nation in the international community (Krasner 1983). Many international regimes begin through a series of accepted norms and practices that countries voluntarily adhere to over time, known as **international customary law**. Although considered legally binding, customary laws are subject to competing interpretations, and according to Marvin Sooros (2002), they tend to be inadequate for addressing complex international environmental problems. The development of formal **international law** is often necessary. In such cases, countries negotiate agreements in the form of **conventions**, **protocols**, and **treaties** that spell out specific expectations, benefits, and ramifications for noncompliance. Such agreements often take decades to negotiate, and in the case of treaties and conventions, require ratification at the nation-state level before they are binding on any of the signing parties. In most cases, **international organizations** (organizations whose members are states that collaborate on specific or multiple issues) have helped facilitate these agreements (see box 6.1 for a list of international organizations important in addressing forest issues).

Although states remain **sovereign** even after signing and ratifying international legal documents, most countries that complete the formal ratification process for treaties and conventions do so with the intention of compliance. States that fail to live up to their international commitments may face difficulty in negotiating future agreements that may be in their interest, and they may also be subject to sanctions from international courts and tribunals. This is why the development of international environmental law

Box 6.1. International Organizations Associated with Forestry Issues

United Nations Environmental Programme (UNEP)
web address: www.unep.org

Commission on Sustainable Development (CSD)
web address: www.un.org/esa/sustdev/csd/csd.htm

Intergovernmental Panel on Forests (IPF) and Intergovernmental Forum on Forests (IFF)
web address: www.un.org/esa/forests/ipf_iff.html

United Nations Food and Agriculture Organization (FAO): Commission on Forest Development in the Tropics
web address: www.fao.org/forestry/index.jsp

World Commission on Forests and Sustainable Development
web address: www.iisd.org/wcfsd/background.htm

Note: List is incomplete and is meant only to highlight some of the most influential international organizations.

is an arduous process of negotiations as all parties engaged in the negotiation process try to formally address a transboundary problem without compromising their national goals and objectives.

In the early stages of formal negotiations, or at times when formal agreements cannot be reached, states may become parties to resolutions (also referred to as "statements of principles" or "plans of action"). Such resolutions are considered "**soft law**" because they represent less formal commitments to addressing the problem than what could be achieved through more formal legal instruments. While such resolutions do not have the strength of treaties or conventions, they often become the cornerstones for more formal international regimes.

The increasing recognition of the importance of tropical forests for **biodiversity** and the atmosphere since the 1980s has led to a growing interest in the international community for addressing deforestation rates around the world. A multitude of **nongovernmental organizations** (NGOs)—organizations whose members are normally not governments but private individuals or groups such as advocacy groups and scientific organizations—started focusing on the issue and working to put global deforestation on the global political agenda (see box 6.2 for a list of NGOs active in forest issues). In the period leading up to the United Nations Conference on Environment and Development (UNCED), known as the **Earth Summit of 1992**, the development of a forest convention was considered. Although the lines of disagreement were not uniformly drawn between developing countries on the one side and developed countries on the other, there was substantial disagreement among developed and developing countries as to the specifics of such an agreement.

During the negotiations for a forest convention, many developing countries refused to take part in the development of a deforestation treaty due to fears of intrusion into areas of national sovereignty and concerns that such an agreement could be used as a **protectionist measure** in trade policy. They emphasized the idea of forests as "national" resources and rejected the competing idea of forests as the "common heritage of mankind," put forth by many advocates of forest preservation (Smouts 2003). Developing countries also pointed out the perceived double standard of developed countries that expressed a desire to protect tropical forests but wanted temperate and boreal forests left out of the negotiations.

Scholars also note that financing was a source of controversy, as even those developed countries supportive of a forest convention were unwilling to provide additional finances to developing countries. Developing countries demanded the creation of a global forest fund to assist in conservation. Many developed countries refused, stating that such funding already existed but was being spent ineffectively and inappropriately by developing countries. In the end, the promise of developing a formal treaty to oversee the sustainable use and conservation of the world's forests was shelved and replaced by a non–legally binding statement of **Forest Principles** for

Box 6.2. Nongovernmental Organizations Active in Forestry Issues

Forest Stewardship Council (FSC)
web address: www.fscus.org

World Wildlife Fund (WWF)
web address: www.worldwildlife.org

Friends of the Earth
web address: www.foei.org

International Institute for Environment and Development (IIED)
web address: www.iied.org

Global Forest Watch
web address: www.globalforestwatch.org/english/index.htm

World Resources Institute
web address: www.wri.org

World Rainforest Movement
web address: www.wrm.org.uy/index.html

International Alliance of Indigenous and Tribal People of Tropical Forests (IAITTF)
web address: www.international-alliance.org

Global Forest Coalition
web address: www.wrm.org.uy/GFC/

Environmental Investigation Agency
web address: www.eia-international.org

Note: This is only a partial list of nongovernmental organizations active in tropical deforestation issues.

management, conservation, and sustainable development of the world's forests, which was signed at the Earth Summit.

In the eyes of political scientists, the lack of a forest convention does not mean that no political or legal regulation of forests exists. On the contrary, regime theorists argue that a global forest regime can be identified. Using Krasner's definition that a regime is a set of "implicit or explicit principles, norms, rules, and decision-making procedures around which actors' expectations converge in a given area" (Krasner 1983, 2), regime theorists point out that a corresponding governance system composed of existing international law elements as well as soft law components and institutions exists for addressing tropical deforestation.

In addition to the Forest Principles adopted at the Earth Summit, the Commission on Sustainable Development (CSD), also formed at the Earth Summit, created an open-ended, ad hoc Intergovernmental Panel of Forests (IPF), which met between 1995 and 1997 to identify strategies for the management, conservation, and sustainable development of all types of forest. The work of the IPF was subsequently continued in the Intergovernmental Forum on Forests (IFF) under the aegis of the United Nations General Assembly Special Session, with the aim of developing a basis for international forest governance. While the IPF and the IFF both developed proposals for action, they failed to achieve consensus on or restart negotiations for a global forest convention. Governments have moved forward, however, to create the United Nations Forum on Forests (UNFF) within the United Nations Economic and Social Council (ECOSOC) as a forum for dialogue and continued policy development, and to foster the implementation of the IPF and IFF proposals for action. In addition, there are also a number of conventions and agreements that cover issues relevant to forests in some form. Box 6.3 provides a list of many of these treaties and conventions along with website addresses.

Next to these various layers of relevant hard and soft institutions, scholars also identify a normative consensus on core elements of a global forest regime (Brunée and Nollkaemper 1996). Specifically, they recognize four common normative elements in international forest agreements today: 1)

Box 6.3. Important International Treaties and Conventions That Address Deforestation

African Convention on the Conservation of Nature and Natural Resources (1968)
Includes a provision to set aside areas for forest reserves and carry out afforestation programs where necessary. Thirteen African countries are signatories to the convention. While important, in the years since the treaty went into effect, many of the countries that are signatories to the convention have suffered from extreme poverty, political upheaval, and war, making it difficult to implement the treaty.

International Tropical Timber Agreement (1983 and 1994)
While stressing the need for sustainably managed forests, this agreement essentially promotes the international timber trade. Currently, 19 countries have ratified the agreement, but fewer than half of the countries that have ratified the agreement are tropical timber producers. The treaty is set up to provide producers and consumers of tropical timber with equal votes.

Convention on Biological Diversity (1992)
Comprehensive treaty that addresses the conservation of biological diversity along with sustainable use and equitable sharing of the benefits of biological diversity. In 2002, specific forest provisions were added that include efforts to reduce threats to forest ecosystems through restoration, agroforestry, watershed management, and protected areas. The convention also has provisions for addressing socioeconomic factors that contribute to deforestation, and assessment and monitoring provisions. The convention was signed by 188 countries, and 168 have since ratified it.

United Nations Convention on Climate Change (UNCCC 1992) **and Kyoto Protocol** (1997)
A key component of this international agreement is the management of the world's forest resources for their value as carbon sinks. Specific language in the agreements addresses reduction in forest degradation and the promotion of carbon sequestration projects using natural, mixed, and plantation forests.

an ecosystem approach stressing in situ conservation, 2) an emphasis on local knowledge and indigenous practices, 3) an emphasis on the principle of participation, and 4) an emphasis on the principle of protected areas (Humphreys 2003). Others highlight progress toward consensus on sustainable forestry through agreements associated with the International Tropical Timber Organization (ITTO), the European Union (EU), and the Montreal Process (for temperate and boreal forests). These agreements stress conservation of biological diversity; maintenance of the productive capacity of forest ecosystems; maintenance of forest ecosystem health; conservation and maintenance of soil and water resources; maintenance of forest contribution to global carbon cycles; maintenance and enhancement of long-term multiple social and economic benefits; and legal, institutional, and economic frameworks for forest management (McDonald and Lane 2004, 64).

Yet many forest conservation advocates and scholars question whether the global community is making real progress on this complex environmental problem. As one scholar notes, "Every single international initiative on forests produced a last-minute agreement to keep talking" (Dimitrov 2003, 136). To date, international regulation of tropical forests remains a hodgepodge of overlapping and autonomous agreements frustrated by national policies.

Private Authority in Global Forest Governance: Promises and Pitfalls

In the last decade, **regime theory** has come to be complemented and to some extent replaced by the **global governance** approach in international relations research. The global governance approach goes further and identifies nonstate actors as pivotal influences in global politics. Advocates of this approach argue that politics in a globalizing world cannot solely be characterized by interaction between states. Rather, nonstate as well as suprastate actors are increasingly acquiring active, and to some extent autonomous, political decision-making capacity (Messner und Nuscheler 1996b, 2003). Scholars note examples of global problem solving and rule setting that are taking place even though a global government does not exist (Rosenau and Czempiel, 1992). In particular, these scholars highlight new opportunities for business and nongovernmental organizations (NGOs) in the design, implementation, and enforcement of standards and regulations for goods and services throughout the global community. Based on this observation, scholars are noting the emergence of "private authority" in global politics. It is this focus on private governance and authority, covered below, that offers particularly fruitful insights into the politics of tropical deforestation today.

The most prominent tools for the exercise of private authority in global governance are self-regulation and **public–private** or **private–private partnerships** (PPPs). Self-regulation refers to the design, implementation, and enforcement of rules and regulations for business by business. PPPs draw on the cooperation between business and governmental actors or NGOs (or even between all three groups of actors in tripartite governance institutions) in the development, implementation, and enforcement of rules and regulations. These governance institutions allow nonstate actors (such as NGOS, corporations, or businesses) not only to influence public political agendas and rules, but to directly set rules and regulations themselves. In the context of the governance of tropical forests, both PPPs and self-regulatory arrangements play a crucial role.

The **Forest Stewardship Council** (FSC) is one of the most important institutions to be mentioned in the context of PPPs (Domask 2003). After an unsuccessful campaign that targeted the International Tropical Timber Organization for a global certification program that would certify tropical timber from sustainable sources, NGOs—most notably the World Wildlife Fund (WWF)—and timber traders created the FSC in the early 1990s. Today, FSC membership consists of a variety of interests, including environmental organizations, foresters, timber traders, indigenous peoples' organizations, community forest groups, and forest product certification organizations from more than two dozen countries, with a balance in membership of developed and developing country organizations.

The FSC's mandate is to promote environmentally appropriate, socially beneficial, and economically viable management of the world's forests. In pursuit of this objective, it has developed a **certification scheme** for "well-managed forests," according to which more than 25 million hectares of forests in 54 countries had been certified by 2002 (box 6.4 details the ten principles of FSC certification). To many observers, the creation of the FSC and the development of its certification scheme promises major progress in protection of tropical forests due to the latter's ability to draw on market power, specifically consumer power, in the diffusion of sustainable forestry practices. With the increasing relevance of global markets and the growing export orientation of developing countries, the possibility of channeling consumer power through large retailers in Organization for Economic Cooperation and Development (OECD) countries could lead to certification as a powerful instrument for the promotion of sustainable forestry.

Under the FSC certification scheme, certification involves an assessment of the original forest site to ensure that the site is managed in a manner consistent with resource sustainability and the maintenance of natural ecosystem services. Since there are differences in these core components, certification divides forest resources into three categories: those from natural forests, plantation forests, and mixed forests. Certification also involves monitoring the chain of custody for timber to ensure that what eventually ends up with a certification label available for consumer purchase is the same forest product that left a certified site.

Next to the FSC, a number of alternative certification schemes—some industry led, and some state led—have been developed in various countries. Examples include the Sustainable Forestry Initiative of the American Pulp and Paper Association, the Pan European Forest Certification System, and the Canadian Standards Association Program by the Canadian Sustainable Forestry Certification Coalition.

Box 6.4. The FSC's Principles of Responsible Forest Management

Principle 1—Compliance with laws and FSC principles
Forest management shall respect all applicable laws of the country in which they occur, and international treaties and agreements to which the country is a signatory, and comply with all FSC Principles and Criteria.

Principle 2—Tenure and use rights and responsibilities
Long-term tenure and use rights to the land and forest resources shall be clearly defined, documented, and legally established.

Principle 3—Indigenous people's rights
The legal and customary rights of indigenous peoples to own, use, and manage their lands, territories, and resources shall be recognized and respected.

Principle 4—Community relations and workers' rights
Forest management operations shall maintain or enhance the long-term social and economic well-being of forest workers and local communities.

Principle 5—Benefits from the forest
Forest management operations shall encourage the efficient use of the forest's multiple products and services to ensure economic viability and a wide range of environmental and social benefits.

Principle 6—Environmental impact
Forest management shall conserve biological diversity and its associated values, water resources, soils, and unique and fragile ecosystems and landscapes, and, by so doing, maintain the ecological functions and the integrity of the forest.

Principle 7—Management plan
A management plan—appropriate to the scale and intensity of the operations—shall be written, implemented, and kept up to date. The long-term objectives of management, and the means of achieving them, shall be clearly stated.

Principle 8—Monitoring and assessment
Monitoring shall be conducted—appropriate to the scale and intensity of forest management—to assess the condition of the forest, yields of forest prod-

ucts, chain of custody, management activities, and their social and environmental impacts.

Principle 9—Maintenance of high conservation value forests
Management activities in high conservation value forests shall maintain or enhance the attributes which define such forests. Decisions regarding high conservation value forests shall always be considered in the context of a precautionary approach.

Principle 10—Plantations
Plantations shall be planned and managed in accordance with Principles and Criteria 1–9, and Principle 10 and its Criteria. While plantations can provide an array of social and economic benefits, and can contribute to satisfying the world's needs for forest products, they should complement the management of, reduce pressures on, and promote the restoration and conservation of natural forests.

Source: www.fsc.org

Companies are also using environmental management systems or codes of conduct as self-regulatory instruments. One example is the scheme developed by the International Standards Organization (ISO) that certifies whether a company's forest management system is likely to meet its specified environmental goals (often referred to as the **ISO 14000 standard** or the ISO 14001 standard). A final example of the influence of PPPs is the World Bank's cooperation with the World Wildlife Fund in the creation of a worldwide system of protected areas.

In sum, private governance institutions increasingly are setting rules and regulations for management and harvesting practices in tropical forests. From the perspective of a political scientist, a number of pivotal questions arise from this observation. Why can private actors develop and enforce such rules and regulations and thus carry out governance functions traditionally considered the domain of the state? Why can business, in particular, obtain the position of a "regulator" of business activities? Furthermore, why would nonstate actors—again, in particular, business—want to do so? And, what is the likely impact of these private governance institutions? Let me discuss the answers to these questions in turn.

Institutions of private governance, such as the FSC or alternative labeling and certification schemes, are spreading because private actors, specifically business and NGOs, have gained authority—that is, "decision-making power over an issue area that is generally regarded as legitimate by participants" (Cutler et al. 1999, 362). Private actors increasingly are acquiring a position in the political arena that infuses their relationships with other actors with an "obligatory quality" (1999, 362). The sources of this new political authority of business and NGOs have pragmatic and moral legitimacy. **Pragmatic legitimacy** is based on the ability to provide desired results rather than on traditional notions of participatory democratic norms and procedures—a source of legitimacy that has become extremely important for business. **Moral legitimacy**, based on perceived notions of fairness and justice, has been a primary source of authority for NGOs, although business is venturing into this territory as well (Fuchs 2005).

Private governance mechanisms derive legitimacy from their fit with various layers of dominant societal norms and ideas. These norms and ideas at the most fundamental level emphasize the benefits of individualism, **decentralization**, and voluntary cooperation. They also stress the superiority of market mechanisms over government intervention (Falk 1999). At a more specific level, they are represented in existing international economic, environmental, and social governance institutions and in the emphasis placed on international trade and economic development objectives (Bernstein and Cashore 2004; Humphreys 2003). As Bernstein (2001) points out, certification and labeling schemes are very compatible with the existing broader norms of **liberal environmentalism**. It is the prevalence of these norms that we recognize in the increasing interest in market-based policy instruments since the 1980s and in the increasing acquisition of authority by private governance institutions thereafter.

But why would business actors want to regulate their own conduct or cooperate with NGOs in regulating this conduct? Adopting a slightly optimistic lens, one could argue that business actors participated because they—like environmental NGOs—want to reduce tropical deforestation and create a level playing field allowing them to do so. Moreover, the regulations may not hurt profits and may even benefit them. Thus, companies may be able to maintain profit levels or even increase them if they can credibly distinguish sustainably harvested timber from conventionally harvested timber via certification and labeling schemes supported by consumers. Finally, business actors may have recognized that their long-term economic prospects, as well as their societal acceptance, depend on improvements in their own conduct.

A more pessimistic view on business's regulation of business activities that always comes up in evaluations of private environmental governance is the threat of **greenwashing** activities. In this case, business self-regulation or PPPs may serve to improve their images through the promotion of their product as certified without substantial improvements in conduct. For instance, a World Wildlife Fund study in 1991 showed that only 3 of 80 self-advertisements by timber importing and retailing companies that claimed that their practices were "ecologically sound" or that their timber was "from sustained yield production" could be linked to actual corresponding efforts. This perspective also corresponds with Dauvergne's (2004) observation that "loggers gain some legitimacy from participation and compromise in high-level meetings, while continuing to log frantically on the ground, paying off powerful state and local allies to maintain access for as long as possible, and distributing largesse along the chain of timber production" (192). Skeptics also believe that many businesses may promote voluntary, weak private regulation simply to prevent more stringent and mandatory public regulation.

Such a critical view, however, may be open to challenges when it comes to the FSC certification scheme. After all, the FSC was created precisely because public actors failed to provide an appropriate framework for forest governance. NGO participation in the FSC would appear to ensure a certain level of stringency in FSC standards. While some environmental groups do argue that these standards are too vague to prevent clear-cutting, ensure the sustainability of harvesting practices, and protect indigenous communities (von Mirbach 1999), the FSC scheme has been perceived by many as holding great potential to fill the gaps in current international law associated with forests. The FSC appears to allow the simultaneous promotion of the interests of various stakeholders and to foster substantial improvements in environmental and social regulatory frameworks effective for forest resources (Gulbrandsen 2004). In the eyes of practitioners, the FSC promises to create effective control and compliance mechanisms for forestry practices through independent third-party auditing and certification schemes and thus to facilitate the influence of market power on business's environmental conduct.

Yet, in the end, the fate of the FSC does lend some credence to a critical perspective on private governance institutions. Since the development of its certification scheme, amendments regarding rule making have been introduced, and substantial variations in regional and local implementation have appeared. More importantly, a number of alternative schemes have been created and paid for by national pulp and paper industries (Gale 2002). These schemes, which generally are considered weaker, are being promoted at the expense of FSC certification. These alternative schemes tend to emphasize process over substance, treating sustainability as discretionary and flexible, a weakness typical of many private environmental and social governance systems (Cashore et al. 2004). While the FSC has been relatively successful in promoting **ecolabels** and certification schemes to promote sustainable consumption and production through scientific and representative accountability, transparency, and equality (Gale 2002), the FSC label has been marginalized in important consumer markets such as the United States and Germany by less stringent schemes (Cashore et al. 2004; Gulbrandsen 2004).

The assumption underlying certification and labeling schemes as governance institutions is that informed consumers will make appropriate purchasing choices. After all, one of the major problems consumers face today is the inability to make consumption decisions based on information regarding the environmental and social costs associated with their choices (Princen 1997). The theory is that certification would help bridge the information gap between production and consumption and thus restore consumer sovereignty.

Unfortunately, in practice, consumers are not as politically aware and active as frequently assumed. As studies have shown again and again, surveys of consumer willingness to pursue environmental consumerism tend to greatly overestimate the role that environmental factors really play in the bulk of everyday consumption decisions (Fuchs and Lorek 2002, 2005). More importantly, environmental consumerism requires that credible and easily accessible information on the environmental characteristics of a product is available. This is the task that environmental labels and certificates can fulfill. Their value, however, is drastically reduced if competing labels with substantially different values enter the market. The majority of consumers will not invest substantial time and energy in distinguishing between more or less valuable labels. Thus, a strategy that some business actors have successfully practiced in a variety of policy fields is to create alternative labels based upon less stringent performance criteria than those created by NGOs. Gale (2002) suggests a simple rule of thumb in this context: consumers should "purchase goods certified by schemes endorsed by reputable environmental organizations, and be skeptical of industry- and government-sponsored logos" (296). However, even such a strategy would require substantial effort on the part of consumers to be able to recognize and distinguish the relevant labels for the different types of product groups and consumption decisions.

In such a situation, NGOs are in a difficult position. First, they frequently cannot compete with business self-advertisements in volume and reach. Secondly, if NGOs highlight the inaccuracy of some labels, the consumer may react by completely disregarding all labels. Thus, such communication has to be crafted and distributed very carefully. This is precisely the situation in which the FSC and environmental organizations aiming to use consumer power to reduce tropical deforestation find themselves today. A large number of consumers in Europe and the United States are aware that it is sensible to look for ecolabels when purchasing timber or products made of tropical wood. Next to the FSC label, however, an increasing number of national industry-sponsored or even firm-specific labels exist, which frequently use lower standards.

At the present time, 4% of global forests are FSC certified as "well managed." Commitments for 20% of wood sold in the U.S. home-remodeling market and more than 20% in the EU market exist (Domask 2003). While these numbers may lead some observers to applaud, they also mean that 96% if the world's forests remain uncertified, and almost 80% of wood sold in the United States and the European Union (not to speak of the Asian and non-OECD markets, where certification has yet to play a major role) will come from uncertified forests. Given the intentional marginalization of the FSC by powerful economic interests in crucial markets, the future diffusion of FSC certification is likely to be slower and less widespread than FSC supporters have hoped. In consequence, FSC certification can play only a complementary role in the protection of tropical forests and cannot alone resolve deforestation problems.

This development is an unfortunate reminder of the tension between two requirements for the success of certification and labeling schemes (Gulbrandsen 2004). On the one side, such schemes need sufficiently stringent environmental standards. On the other side, they require sufficiently widespread participation of producers. Consumers may, of course, create pressure for broad participation, but generally they lack the capacity to do so. The reality is that stringent performance standards rarely find support in the business community. In the case of the FSC, the compromise between NGOs and business actors may have been possible partly because social, environmental, and economic actors of varying sizes had equal influence (in fact, the large timber companies were somewhat critical of the FSC from the beginning because they felt underrepresented). But this is often not the case in other instances of international or domestic policy formation.

Importantly, FSC certification can play only a complementary role for other reasons as well. It is important to keep in mind, for instance, that a large share of certified forests are plantations, which lack many of the ecological benefits of old-growth forests and whose impact on the fate of old-growth forests is, in fact, unclear (von Mirbach 1999). Moreover, certification still falls short of a full internationalization of environmental and social costs and thus does not sufficiently change the land use decisions individuals make. Industrial logging, which can be influenced by certification, is a major source of tropical deforestation in some areas and regions. In other areas, however, agricultural conversion and fuelwood use, on which certification has no impact, are the strongest driving forces (von Mirbach 1999).

We should not reject private governance efforts with respect to tropical deforestation too easily, though, especially not the FSC efforts. After all, prior to the establishment of the FSC, governments had demonstrated neither willingness nor capacity to create adequate rules and regulations. At the international level, few governments had even shown interest in participating in negotiations for a forest convention in preparation of the Earth Summit (Kolk 1996). At the national level, misguided governmental policies as well as corruption were among the major drivers of deforestation (in part because of the influence of large logging and timber trading companies). Moreover, even those governments willing to design and implement appropriate regulations tend to face severe capacity problems when it comes to enforcement. Thus, private governance schemes with respect to forests may be better than not doing anything at all. In their current form, though, this is only the case as long as they are not interpreted as sufficient mechanisms of governance.

Finally, one should be aware that the future of such private governance efforts themselves is uncertain. As pointed out above, they derive their authority from a focus on the legitimacy of outputs and from the assumption that these private governance schemes will provide results. Because the results themselves are open to interpretation, some level of credible achievement will be needed for this legitimacy to last.

Conclusions

What do the arguments and findings presented in this chapter mean in terms of future policy and research needs? On the policy side, it should be clear that the protection of tropical forests requires governance institutions that go beyond current certification and labeling schemes. Either the existing private governance institutions need to be strengthened in terms of their reach and effectiveness, or international law needs to fill the existing gap in governance. Unfortunately, the distribution of interests and the balance of power among relevant actors suggests that neither development is likely.

Scholars and activists have placed their hopes for democratic and effective global governance of tropical forests on NGOs, yet the actual potential of NGOs to successfully influence governance is limited. NGOs often lack sufficient resources to successfully influence policy processes at both the international and national levels. While they have proven to be effective at the agenda-setting stages, they tend to lose influence in the long and drawn-out policy processes necessary for establishing formal policy. This weakness of NGOs is further enhanced by the lack of cohesion (differing goals, strategies, and objectives) among NGOs in this policy arena. Likewise, NGO resources are often too scarce to allow for effective monitoring of existing policy.

On the part of governments, there is little indication of an overall change in attitudes and practices regarding the global management of tropical forests. At the international level, there continues to be no agreement on a forest convention on the horizon. At the national and subnational levels, misdirected policies, an acceptance of compliance failure, and practices of corruption and political **clientelism** continue to exist, thereby diluting and undermining positive efforts at the international level (Dauvergne 2004). Even governments aiming to foster sustainable forestry practices in their countries face severe limits in resources vis-à-vis highly mobile logging firms, with intricate layers and webs of firms allowing easy concealment of illegal logging and smuggling (Dauvergne 2004). Finally, national and international norms continue to prioritize free trade and economic growth over environmental and social sustainability.

Thus, one can only suggest that NGOs pursue a mixed strategy. On the one side, they should invest in sophisticated communication strategies to make the FSC label known and distinguishable for consumers. On the other side, they need to improve coalition building among themselves, as well as with those governments inclined to support international forest governance. The latter governments should also be aware of the need for funding NGOs and other governments in developing countries that have a real interest in supporting sustainable forestry.

Change, finally, could also come from the arrival of a new powerful actor on the scene. The expansion of the United Nations Environment Programme (UNEP) into a global environmental organization (GEO), with broad competencies, sanctioning, and enforcement capacities similar to the WTO, would be one possibility in this context. Yet even a strengthened environmental intergovernmental organization (IGO) would still have to deal with governments as well as powerful economic interests opposing international forest governance. Thus, improvements in global forest governance after the creation of a GEO cannot be taken for granted.

In terms of research needs, three areas of inquiry that promise important new insights can be identified. First, the rule-setting activities of private actors deserve more attention, both in the context of deforestation and in general. Given the large potential of private governance and the repeated disappointment in its actual aims and achievements, more information on strategies to improve these aims and achievements is important. In this context, empirical studies with more extensive and systematic comparisons between relevant governance institutions are needed, paying special attention to the distribution of interests and power among the actors involved. Such studies could provide useful insights on the conditions under which private governance institutions are particularly effective and therefore legitimate.

Second, and in relation to the first area of inquiry, the discursive activities of private actors need to be studied more. After all, the competition between different labeling and certification schemes determines their relative success and thereby their potential contribution to the pursuit of environmental and social policy objectives. Such inquiries need to pay attention, for instance, to the place of business in society in general and to the relationships among the media, business, and politics in particular. Some insights in this area can be drawn from scholars and practitioners in marketing, public relations, and communication science. Moreover, political scientists need to examine the context of larger legitimizing ideologies and business's ability to pursue concrete political strategies and proposals.

Third, the interaction between business power and state power, as well as between business power and the power of civil society, in the governance of tropical forests deserves more attention. As the analysis presented here has shown, it is difficult to evaluate the contributions of these actors to current deforestation patterns and processes separately. Corruption and political clientelism require the involvement of both groups of actors in order to exist. Likewise, the potentials and threats of private governance depend upon their being embedded in a public regulatory framework or lack thereof. From this perspective, the issue of collective-action problems and **free riding** among business actors moves into the focus of attention. How can business actors wanting to protect tropical forests through sustainable forestry practices protect themselves against unfair competition and damage to reputations when it is caused by others?

Both policy development and further research are urgently needed. At the present time, most political scientists studying tropical forests tend to have a somewhat pessimistic outlook on their fate. As Humphreys (2003) argues, destructive forces shaped by broader socioeconomic conditions still triumph over sustainability. Currently, broader socioeconomic conditions influence forest management and harvesting decisions made by relevant individuals and groups. Improving the interplay of public and private elements of global forest governance is thereby necessary. As Dauvergne (2004) states, "The process of change is currently far too slow to save the remaining old-growth commercial forests" (193). These trends and developments will only turn around if we make progress in understanding the political and socioeconomic processes and structures that drive tropical deforestation. Saving a significant share of the remaining tropical forests will require the creation and expansion of relevant governance activities.

REFERENCES

Bernstein, S. 2001. *The compromise of liberal environmentalism.* New York: Columbia University Press.

Bernstein, S., and B. Cashore. 2004. The two-level logic of non-state global governance. Paper presented at the Changing Patterns of Authority in the Global Political Economy conference, Tübingen, Germany, October 14–16.

Brunée, J., and A. Nollkaemper. 1996. Between the forests and the trees: An emerging international forest law. *Environmental Conservation* 23(4):307–314.

Cashore, B., G. Auld, and D. Newsom. 2004. *Governing through markets: Forest certification and the emergence of non-state authority.* New Haven, Conn.: Yale University Press.

Cutler, C., V. Haufler, and T. Porter, eds. 1999. *Private authority and international affairs.* Albany: State University of New York Press.

Czempiel, E.-O. 1992. Governance and democratization. In *Governance without government.* Edited by J. Rosenau and E.-O. Czempiel. Cambridge: Cambridge University Press.

Dauvergne, P. 2004. The environmental challenge to loggers in the Asia-Pacific: Corporate practices in informal regimes of governance. In *The business of global environmental governance.* Edited by D. Levy and P. Newell. Cambridge, Mass.: MIT Press.

Dimitrov, R. 2003. Knowledge, power, and interests in environmental regime formation. *International Studies Quarterly* 47(1):123–50.

Domask, J. 2003. From boycotts to global partnership: NGOs, the private sector, and the struggle to protect the world's forests. In *Globalization and NGOs.* Edited by J. Doh and H. Teegen. Westport, Conn.: Praeger.

Easton, D. 1971. *The political system: An inquiry into the state of political science.* New York: Knopf.

Falk, R. 1999. Liberalism at the global level. In *Toward genuine global governance.* Edited by E. Harris and J. Yunker. Westport, Conn.: Praeger.

Fuchs, D. 2005. *Understanding business power in global governance.* Baden-Baden, Germany: Nomos.

Fuchs, D., and S. Lorek. 2002. Globalization and sustainable consumption. *Global environmental politics* 2(1):19–45.

———. 2005. The implications of the global political setting and economic concerns for consumption governance. *Journal of Consumer Policy* 28(3):261–88.

Gale, F. 2002. Caveat certificatum: The case of forest certification. In *Confronting consumption.* Edited by T. Princen, M. Maniates, and K. Conca. Cambridge, Mass.: MIT Press.

Gulbrandsen, L. 2004. Overlapping public and private governance: Can forest certification fill the gaps in the global forest regime? *Global Environmental Politics* 4(2):75–99.

Humphreys, D. 2003. Life protective or carcinogenic challenge? Global forests governance under advanced capitalism. *Global Environmental Governance* 3(2):40–55.

———. 2004. Redefining the issues: NGO influence on international forest negotiations. *Global Environmental Politics* 4(2):51–74.

Kolk, A. 1996. *Forests in international environmental politics: International organizations, NGOs and the Brazilian Amazon.* Utrecht, the Netherlands: International Books.

Krasner, S. 1983. *International regimes.* Ithaca, N.Y.: Cornell University Press.

Lasswell, H. 1958. *Politics: Who gets what, when, how.* New York: Meridian Books.

Lasswell, H., and A. Kaplan. 1950. *Power and society: A framework for political inquiry.* New Haven, Conn.: Yale University Press.

McDonald, G. T., and M. B. Lane. 2004. Converging global indicators for sustainable forest management. *Forest Policy and Economics* 6(1):63–70.

Messner, D., and F. Nuscheler. 2003. Das Konzept Global Governance: Stand und Perspektiven. INEF Report 67. Duisburg: Institut für Entwicklung und Frieden.

———. 1996. Global Governance: Organisations elemente und Säuleneiner Weltordnungspolitik. In *Weltkonferenzen und Weltberichte,* ed. D. Messner and F. Nuscheler. Bonn: Dietz.

Princen, T. 1997. The shading and distancing of commerce: When internalization is not enough. *Ecological Economics* 20:235–53.

Rosenau, J., and E.-O. Czempiel, eds. 1992. *Governance without government.* Cambridge: Cambridge University Press.

Smouts, M.-C. 2003. *Tropical forests, international jungle.* New York: Palgrave Macmillan.

Sooros, M. 2002. Negotiating our climate. In *Global Climate Change.* Edited by S. Spray and K. McGlothlin. Lanham, Md.: Rowman & Littlefield Publishers.

von Mirbach, M. 1999. Demanding good wood. In *Voluntary initiatives: The new politics of corporate greening.* Edited by R. Gibson. Peterborough, Ontario: Broadview Press.

SUGGESTED READINGS

Barbosa, L. 2000. *The Brazilian Amazon rainforest: Global ecopolitics, development and democracy.* Lanham, Md.: University Press of America.

Gale, F. 1998. *The tropical timber trade regime.* London: Macmillan.

Gibson, R., ed. 1999. *Voluntary initiatives.* Peterborough, Ontario: Broadview Press.

Haufler, V. 2001. *A public role for the private sector.* Washington, D.C.: Carnegie Endowment for International Peace.

Hirsch, P., and C. Warren, eds. 1998. *The politics of environment in Southeast Asia: Resources and resistance.* London: Routledge.

Humphreys, D. 2003. United Nations Forum on Forests. *Global Environmental Change* 13(4):319–23.

Levy, D., and P. Newell, eds. 2004. *The business of environmental governance.* Cambridge, Mass.: MIT Press.

Poore, D. 2003. *Changing landscapes: The development of the International Tropical Timber Organization and its influence on tropical forest Management.* London: Earthscan.

Rosendal, G. K. 2001. Overlapping international regimes: The case of the Intergovernmental Forum on Forests (IFF) between climate change and biodiversity. *International Environmental Agreements: Politics, Law, and Economics* 1(4):447–68.

Tarasofsky, R., ed. 1999. *Assessing the International Forest Regime.* Cambridge: IUCN.

CHAPTER 7

Concluding Thoughts

Sharon L. Spray
and
Matthew D. Moran

Trying to predict the future of tropical forests, like predicting any future event, is risky business. Economic situations, political interactions, and societal values will all play a role in the future of the tropics. Moreover, different parts of the world are likely to respond in different ways to the problems of tropical **deforestation**. Therefore, one can imagine certain geographic areas maintaining viable tropical forests while other areas show complete elimination. Other areas may show "recovery" and return to some resemblance of a functional forest.

One possibility is "business as usual," in which case tropical forests will continue to be lost or altered at about the same rate as today. If this process continues for the next 50 years, very little tropical forest will remain in a natural state, and much will be completely lost. The world will undoubtedly lose many species as the altered habitats that replace tropical forests will support far fewer types of organisms. Many surviving tropical species will be relegated to the few nature preserves that are well protected. The continued loss of forest will probably cause numerous sociopolitical problems as scarcity of wood, altered climate, and land and water shortages all act to increase the challenges to tropical countries. Some countries that have suffered severe forest loss (e.g., Haiti) already show these types of problems.

It is probably not realistic, though, to predict that current conditions will continue in all parts of the world. Several Central American countries, especially Costa Rica, Panama, Belize, and to a lesser extent Mexico and Nicaragua, have shown an increasing interest in conservation of tropical forests. In some cases, this has been economically beneficial to these countries through international aid and tourism, and perhaps by focusing the economy on more productive economic sectors. However, most other parts of the world have shown little or no progress in addressing the issue of tropical deforestation. The largest remaining tropical forest, the Amazon basin, continues to shrink around the edges at an alarming rate, with an ever-increasing road development system fragmenting the forest and increasing migration and extraction within it. Although Brazil has made several attempts to address deforestation in the Amazon, little real progress has been made on the ground.

The trend in central Africa is also not positive. While some African countries, such as Gabon—which recently made the commitment to set aside 10% of its land mass for parks, reserves, and sanctuaries—have shown increasing interest in preservation of their natural wealth, deforestation is escalating elsewhere on the continent, with no easy solution. The complicated mix of variables in Africa includes poverty, increased resource extraction tied to economic structural adjustment programs, distorted tax incentive policies, currency devaluations, and war. The general lack of political stability in many central African countries has made it very difficult to address deforestation problems, and there is no quick solution to current trends.

Southeast Asia, home to the highest diversity of forests, is currently suffering extremely high rates of deforestation. Much of this is due to rapid economic expansion, human population growth, trade activity such as the export of cash crops (e.g., palm oil), and corruption. The future of these forests and of many of the charismatic animals that inhabit them, such as the Orangutan (*Pongo pygmaeus*), is bleak.

Ultimately, the question is whether we can limit and reverse tropical deforestation and at the same time develop viable and rewarding lives for the people in countries that contain tropical forests. Both are imperatives. Almost all countries that contain significant areas of tropical forests also contain large populations of desperately poor people, many of whom depend upon forest resources for their daily needs, including fuelwood and food. Therefore, traditional conservation measures such as the creation of national parks and preserves (which in many cases require displacement of forest-dwelling people), along with sustainable forestry practices and pollution control measures, will be inadequate to stop tropical forest destruction. The notion of creating parks is itself a controversial topic. Well-protected parks have undoubtedly helped preserve some tropical forests. For example, the Monte Verde Preserve in Costa Rica and Manu National Park in Peru each protect high-quality tropical forest. They also receive substantial international support and lucrative tourist visitation. However, many protected areas are little more than paper parks, suffering from uncontrolled settlement, illegal logging, and rampant hunting of animals. Additionally, the establishment of parks and preserves seldom addresses the needs of local people.

In the last decade, global and domestic indigenous rights groups have emerged as significant political forces in tropical countries. Groups such as the International Alliance of Indigenous and Tribal People of Tropical Forests (IAITTF), Friends of the Earth, and the Global Forest Coalition have lobbied the United Nations for an international agreement to honor the rights of indigenous communities. While the issue of indigenous rights is viewed as a human rights issue in many regions of the world, in the tropics such rights are also promoted as a means for better forest management.

In cases throughout the tropics, indigenous peoples have been displaced through government-supported land conversion programs (timber concessions; the creation of oil, palm, and pulp plantations; the expansion of the cattle industry) and illegal land use (illegal timber harvesting and illegal expansion of agricultural lands). Not only has this displacement increased rural poverty and shifted migration patterns into other parts of tropical forests, but activists suggest that much of the indigenous knowledge associated with sustainable use of forest resources is being lost. Activists in tropical countries are pressuring governments for (at the most) greater community-based control of forest resources and (at the least) meaningful community input into forest management schemes and forest policy.

It should be noted that shifting control of forests to community control is not the same as "decentralized authority" over forests. Many countries, including Indonesia and Brazil, have moved toward decentralized decision making regarding forest policy, but this is not the same as local *community* control that accords customary land tenure rights to indigenous groups. In the case of decentralized decision making, policy formulation and government decisions concerning forest resources shift from the national-government level to more local levels of government, but this does not necessarily mean that there is community decision making over forest resources or that there is a guarantee for better forest management (Kunanayagan 1998). As researcher Chris Bennett suggests,

> Sustaining natural resources to reflect their true economic and social value is often, at best, a secondary consideration of local government. Furthermore, rights to natural resources may be captured by local elites consisting of senior bureaucrats and favored businessmen, typically excluding the poor, the majority of whom are heavily dependent upon natural resources. (2002)

Even when local government officials are well intentioned, "the failure to draw upon the knowledge of villagers through meaningful participation inhibits the development of sound policy formation" (Kunanayagan 1998, 140). Indigenous rights activists want control of forest resources based on the rights of communities as a whole (customary land tenure rights and practices) rather than policies that favor individual property rights or direct government management.

Several tropical nations, such as Papua New Guinea, Uganda, and the Philippines, have already adopted formal policies that respect customary land tenure rights, making it possible for community-based ownership of forest tracts, but the success of these policies is still unknown. What is known is that in many parts of the tropics, the struggle for control of forest resources is violent and is predicated on widespread government corruption and illegal activity that must be addressed before deforestation trends can be reversed.

Kenya is one such country struggling to preserve its limited tropical forests in the face of corruption. One study found that over 90% of all interactions with government officials involved bribery or some other form of corruption (*The Economist* 2004). Kenyan Wildlife Service officials have often been involved in poaching, logging, and other destructive activities within national parks. Unfortunately, this example is not unique. The nongovernmental organization Environmental Investigation Agency estimates that in Indonesia's lowland tropical forests, there are twice as many illegal sawmills as legal ones, and illegal timber extraction from national parks and other areas accounts for approximately 70% of Indonesia's log production (*International Herald Tribune*, August 26, 2002). Changing the long-standing political cultures of concentrated power and corruption so common in many countries with tropical forest resources will not be easy.

Many scholars studying deforestation suggest that increased levels of democracy in tropical countries may hold the answer to greater preservation of forest resources. Rodger Payne (1995) explains that the connection between democracy and environmental protection rests on five basic arguments. First, democracies are rights-based political systems in which citizens can lobby their governments individually or collectively to protect the environment. In many of the countries where deforestation continues at alarming rates, open opposition to government policy is pursued at great personal risk to activists. For citizens to have greater input into the use of forest resources, they must be accorded the rights to petition government and to protest and agitate for change (Payne 1995). Coinciding with the rights of individuals to speak freely is the right of media to report, without fear of retribution, the activities of government and the state of environmental quality.

The second argument favoring democratic reforms in tropical countries is that democratic governments must be responsive to public opinion or risk losing their electoral support. In this sense, grassroots organizations in open societies play an important role in counterbalancing the influence of powerful elites and international business interests through public awareness campaigns, lobbying efforts, and the shaping of electoral issues (Payne 1995). These groups at times may even transform into "green" political parties that transform the composition of legislative bodies into more environmentally responsive institutions.

A third argument in favor of democratic reforms is that democratic states tend to adopt more innovative and flexible policy approaches than do their nondemocratic counterparts. This is likely connected to greater dissemination of scientific and policy information, but also to the fact that democratic countries are more likely to interact with other nation-states and learn about their policy successes and failures (Payne 1995). While fair and predictable election cycles may increase government legitimacy, they may also promote more turnover in policymaking positions as new administrations enter and leave government, thereby allowing the entrance of new ideas and approaches to environmental problems.

A fourth argument is associated with the concept of internationalism. Specifically, democratic governments are more likely to take part in international environmental institutions, to abide by environmental agreements, and to respond to international pressure. As Payne suggests, "Global organizations can apply a variety of pressures—some subtle and some direct—to get governments to accept environmental norms" (Payne 1995, 47). Environmental institutions also provide resources to member states, such as shared scientific data, the facilitation of environmental partnerships (such as the joint implementation projects under the Kyoto Protocol), and the transfer of technology. In many cases, once party to international agreements, countries are monitored and pressured by nongovernmental organizations to live up to established laws and principles associated with these agreements. Those countries that do not fulfill their commitments risk a loss of international prestige, possible international sanctions, and increased domestic pressure (Porter et al. 2000).

And, lastly, Payne (1995) points out that all democracies have market-based economic systems. While critics of capitalism may find this last point dubious given the common capitalist market failures such as pollution externalities, overexploitation of limited resources, and so on, Payne suggests that the growth of market-based approaches can also promote environmentally sound business practices associated with consumer pressure for "green" products. He points out that many democratic countries are experiencing growth in environmental technology markets. Payne's market-based argument, however, is sure to bring criticism from many who can cite plenty of anecdotal examples of problems associated with unchecked markets and industries, the concentration of natural resources in public hands—managed by dictatorial, authoritarian, and military regimes—is part of the legacy of corruption and deforestation that has accelerated deforestation in many parts of the world.

According to a report released in 2003 by the World Forests, Society and Environment (WFSE) program, four-fifths of the world's forests are under public ownership. The study documents that where states own most of the timber resources, sales of standing timber—or stumpage sales—are priced lower than in the competitive world market. While it may seem unreasonable for governments to sell timber resources for far less than the world market value, logging concessions are used in many areas of the tropics as a means to solidify political support. Government officials provide sweetheart deals to military leaders and well-connected elites who can make high profits from timber sales or from the conversion of land to other uses and who in turn support the existing regime. The most blatant examples of the use of timber concessions for political gain occurred under dictatorships such as those of Marcos in the Philippines and Suharto in Indonesia, and currently under the military leaders of Myanmar (formerly Burma). Nevertheless, the practice remains common in many parts of the tropics, including Central and South America (Faber 2001).

Throughout this volume, authors have stressed that the variables that cause tropical deforestation vary from region to region, with some variables playing little or no role in some regions and larger roles in others. One should therefore be careful not to assume that political reforms of any kind alone can address tropical deforestation. Political reform must be accompanied by other social reforms. Evidence of this showed up in a 1998 study by Manus I. Midlarsky. Using World Resources Institute (WRI) deforestation rates data from 1981 to 1990, Midlarsky found a collinear relationship between levels of democracy and rates of deforestation (the greater the level of democracy, the greater the rate of deforestation). Midlarsky speculates that this counterintuitive result may reflect two conditions related to deforestation that are not associated with other measures of environmental protection. The first theory is that many of the tropical countries that are experiencing high rates of deforestation also have very high rates of landless peasants that must be addressed for social and political reasons. Democratically elected governments in these areas may fear being ejected from office by these disgruntled, land-poor peasants and are therefore acceding to their demands for greater access to forest lands. In cases where the landless poor do not vote, civil disobedience such as squatting and demonstrations, which are publicized by the freer press that accompanies democratic regimes, may be sufficient for influencing forest policy, especially as these events draw international media attention.

Midlarsky's second theory for explaining why democracy is not sufficient for curtailing deforestation is that not all interests in a democracy are equal. While democratic reforms may allow for greater environmental advocacy, large-scale cattle ranchers, farmers, and multinational corporations still hold substantial political power in newly democratizing nations with tropical forests. We suggest that forest protection is still at odds with the agricultural development that tropical countries continue to promote for export revenues and for feeding growing urban populations.

Even in the face of conflicting data on the relationship between democracy and environmental preservation, we should not easily dismiss the value of many of the components of democratic regimes or the anecdotal evidence that democratic rule is more favorable than authoritarian regimes when it comes to preserving the overall ecological wealth of a nation. Increased democratic rule has paved the way in many countries for the growth of environmental movements and for greater international pressure to pursue stronger environmental policies. We also know, however, that in many areas where tropical forests continue to disappear at alarming rates, political stability—democratic or otherwise—is elusive.

David Kaimowitz, director of the Centre for International Forestry Research in Indonesia, estimates that "well over 50% of the world's tropical forests are in countries that have had violent conflict in the past fifteen years" (*The Economist* 2004). And the reality is that you must have government stability to have government accountability, whether for enforcing existing policy or for generating new policy.

Clearly, tropical countries must develop the means to promote economic development outside of natural resource extraction. To date, development initiatives throughout tropical climates have exacerbated deforestation as pressure for rapid economic growth and foreign capital have opened up forest regions to logging, mining, and agricultural interests. The foreign debt that many developing countries acquired in previous decades has contributed to exploitation of their natural resources to earn the foreign exchange necessary for debt repayment. The need to generate cash has increased the pressure to sell tropical hardwoods, to open forest areas to mining and petroleum production, and to promote the conversion of forest land to agricultural exports like soy, cotton, and cattle (Leigh 1998; Place 2001; Resosudarmo 2002). Yet trade liberalization has in some cases moved economies from natural resource extraction to the production of value-added goods and services. For example, Thailand, Malaysia, and Costa Rica have had rapid growth of industrial economies, which may in part explain their recent concerns with forest protection. As populations become more affluent, the tendency is to increase the intrinsic value of the natural world (Inglehart 1977). Therefore, the movement of economies from agriculture and natural resource utilization to goods and services tends to lead to more forest protection.

As previously mentioned, one nonagricultural industry that shows progress along this front is tourism. Throughout the world, tourism is the fastest-growing industry and one of the largest industries on Earth. In some countries (e.g., Costa Rica), it is already the number-one industry. In many countries, tourism, although not the top industry, is still the biggest source of foreign currency. Yet the effect of tourism on deforestation rates is difficult to predict. It is well accepted that tourism has stopped deforestation in isolated areas. For instance, the cloud forests of Monteverde, Costa Rica, have benefited greatly by the boom in tourism, as it is very likely they would have been cleared for timber and agriculture if it were not for the tourist industry. Basically, the forest in that area is worth more intact than could be made through short-term exploitation. Whether this model can be adopted in other areas, however, is an open question.

Many tropical forests are in remote areas with poor or nonexistent infrastructure, making visits by tourists very difficult. Tourism also contributes to deforestation and pollution through the amenities (roads, lodges, etc.) necessary to support the tourism industry. The question then remains: Is tourism a net positive or net negative effect on tropical forests? It is probably a net positive as some forests are conserved and public awareness is raised. Meanwhile, the destructive aspects of tourism dwarf those of extractive resource use. However, tourism is unlikely to be *the* solution to tropical deforestation. There are only so many tourists, and many tropical forests are vast and are located in remote areas that are unlikely to benefit from the tourist industry. In sum, tourism alone is not the answer.

As we said in the preface of this text, this book is not meant as a policy primer. Rather, this text is designed to provide readers with several disciplinary perspectives helpful in understanding the complexity of variables associated with tropical deforestation. While not advocating specific policy proposals, we do suggest that current patterns of deforestation will likely require region-specific policies that provide strict protection for some forests, multiple use for others, and social and legal reforms in most. Tropical countries will need diversification and modernization of their economies and increased commitments to provide the necessary resources for oversight and maintenance of forest resources. The international community will also need to do its part by strengthening international agreements associated with the preservation of forest resources; by working cooperatively to support certification schemes (e.g., currently no FSC-certified timber comes out of Africa); and by helping developing countries address issues associated with poverty and population growth.

Success is neither hopeless nor guaranteed. Some countries such as Costa Rica and Belize have made great strides toward saving their tropical forests. It is our hope that their successes can be replicated in larger countries that contain more significant areas of forest. The Amazon basin still exists as a relatively intact **ecosystem**, one that teems with incredible life-forms and astounding levels of diversity. Central Africa still contains a large block of forest, and the islands of Borneo and New Guinea support unique and functional forest communities. There is still time to protect much of the world's tropical forests for the future, but time is short.

REFERENCES

Bennett, C. P. A. 2002. Responsibility, accountability, and national unity in village governance. In *Which way forward? People, forests and policymaking in Indonesia.* Edited by C. J. P. Colfer and A. P. Resosudarmo. Washington, D.C.: Resources for the Future Press, 60–80.

Colfer, C. J. P., and A. P. Resosudarmo, eds. 2002. *Which way forward? People, forests and policymaking in Indonesia.* Washington, D.C.: Resources for the Future Press.

Faber, D. 2001. Revolution in the rainforest. In *Tropical rainforests: Latin American nature and society in transition.* Edited by S. E. Place. Wilmington, Del.: Scholarly Resources, 97–122.

Inglehart, R. 1977. *The silent revolution: Changing values and political styles among Western publics.* Princeton, N.J.: Princeton University Press.

Kunanayagam, R., and K. Young. 1998. Mining, environmental impact and dependent communities: The view from below in East Kalimantan. In *The politics of environment in Southeast Asia: Resources and resistance/* Edited by P. Hirsch and C. Warren. London: Routledge, 139–58.

Leigh, M. 1998. Political economy of logging in Sarawak, Malaysia. In *The politics of environment in Southeast Asia: Resources and resistance.* Edited by P. Hirsch and C. Warren. London: Routledge, 93–106.

Lutzenberger, J. A. 2001. Who is destroying the Amazon rainforest? In *Tropical rainforests: Latin American nature and society in transition.* Edited by S. E. Place. Wilmington, Del.: Scholarly Resources, 123–135.

Midlarsky, M. I. 1998. Democracy and the environment: An empirical assessment. *Journal of Peace Research* 35(3):341–61.

Payne, Rodger A. 1995. Freedom and the Environment. *Journal of Democracy* 6(3):41–55.

Place, S., ed. 2001. *Tropical rainforests: Latin American nature and society in transition.* Wilmington, Del.: Scholarly Resources.

Porter, G., J. W. Brown, and P. S. Chasek. 2000. *Global environmental politics.* 3rd ed. Boulder, Colo.: Westview Press.

Resosudarmo, I. A. 2002. Timber management and related policies: A review. In *Which way forward? People, forests and policymaking in Indonesia.* Edited by C. J. P. Colfer and A. P. Resosudarmo. Washington, D.C.: Resources for the Future Press, 161–90.

The Economist. 2004. Where graft is merely rampant. *The Economist* 373:65.

Bailey, S. W., J. W. Hornbeck, C. T. Driscoll, and H. E. Gaudette. 1996. Calcium inputs and transport in a base-poor forest ecosystem as interpreted by Sr isotopes. *Water Resources Research* 32:707–19.

Brady, N. C., and R. R. Weil. 2002. *The nature and property of soils.* 13th ed. Upper Saddle River, N.J.: Prentice Hall.

Bridges, E. M. 1997. *World soils.* 3rd ed. Cambridge, U.K.: Cambridge University Press.

Buol, S. W., P. A. Sanchez, J. M. Kimble, and S. B. Weed. 1990. Predicted impact of climate warming on soil properties and use. In *Impact of carbon dioxide trace gases and climate change on global agriculture.* Edited by B. A. Kimbal et al. ASA Special Publication 53. Madison, Wis.: Soil Science Society of America, 71–82.

Casper, B. B., H. J. Schenk, and R. B. Jackson. 2003. Defining a plant's belowground zone of influence. *Ecology* 84:2313–21.

Crews, T. E., K. Kitayama, J. H. Fownes, R. H. Riley, D. A. Herbert, D. Mueller-Dombois, and P. M. Vitousek. 1995. Changes in soil phosphorus fractions and ecosystem dynamics across a long chronosequence in Hawaii. *Ecology* 76:1407–24.

Cuevas, E. 1995. Biology of the below-ground system of tropical dry forests. In *Seasonally dry tropical forests.* Edited by S. H. Bullock et al. Cambridge, U.K.: Cambridge University Press, 362–83.

Davidson, E. A, L. V. Verchot, J. H. Cattanio, I. L. Ackerman, and J. E. M. Carvalho. 2000. Effects of soil water content on soil respiration in forests and cattle pastures of eastern Amazonia. *Biogeochemistry* 48:53–69.

Ewel, J., C. Berish, and B. Brown. 1981. Slash and burn impacts on a Costa Rican wet forest site. *Ecology* 62:816–29.

Garcia-Montiel, D. C., C. Neill, J. Melillo, S. Thomas, P. A. Steudler, and C. C. Cerri. 2000. Soil phosphorus transformations following forest clearing for pasture in the Brazilian Amazon. *Soil Science Society of America Journal* 64:1792–1804.

Goodland, R., and H. Irwin. 1975. *Amazon jungle: Green hell to red desert?* Amsterdam: Elsevier.

Gregorich, E. G., L. W. Turchenenk, M. R. Carter, and D. A. Angers, eds. 2001. *Soil and environmental science dictionary.* Canadian Society of Soil Science. Boca Raton, Fla.: CRC Press.

Heckenberger, M. J., A. Kuikuro, U. Tabata Kuikuro, J. C. Russell, M. Schmidt, C. Fausto, and B. Franchetto. 2003. Amazonia 1492: Pristine forest or cultural parkland? *Science* 301(5640):1710–14.

Hedin, L. O., P. M. Vitousek, and P. A. Matson. 2003. Nutrient losses over four million years of tropical forest development. *Ecology* 84:2231–55.

Houghton, R. A., D. L. Skole, C. A. Nobre, J. L. Hackler, K. T. Lawrence, and W. H. Chomentowski. 2000. Annual fluxes of carbon from deforestation and regrowth in the Brazilian Amazon. *Nature* 403(20):301–4.

Huntington, T. G. 2000. The potential for calcium depletion in forest ecosystems of southeastern United States: Review and analysis. *Global Biogeochemical Cycles* 14:623–38.

Johnson, C. M., A. H. Johnson, J. Frizano, I. C. Vieria, and D. J. Zarin. 2001. Carbon and nutrient storage in primary and secondary forests in eastern Amazonia. *Forest Ecology and Management* 147(2–3):245–52.

Juo, A. S. R., and A. Manu. 1996. Chemical dynamics in slash and burn agriculture. *Agriculture, Ecosystems and Environment* 58:49–60.

Kauffman, J. B., D. L. Cummings, and D. E. Ward. 1998. Fire in the Brazilian Amazon 2: Biomass, nutrient pools, and losses in cattle pastures. *Oecologia* 104:397–408.

Lal, R. 1990. *Soil erosion in the tropics: Principles and management.* New York: McGraw-Hill.

Lawrence, D., and W. H. Schlesinger. 2001. Changes in soil phosphorus during 200 years of shifting cultivation in Indonesia. *Ecology* 82:2769–80.

Lehmann, J., D. Gunther, M. Socorro da Mota, M. Pereira de Almeida, W. Zech, and K. Kaiser. 2001. Inorganic and organic soil phosphorus and sulfur pools in an Amazonian multistrata agroforestry system. *Agroforestry Systems* 53:113–24.

Maass, J. M. 1995. Conversion of tropical dry forest to pasture and agriculture. In *Seasonally dry tropical forests.* Edited by S. H. Bullock et al. Cambridge, U.K.: Cambridge University Press, 399–422.

Margolis, H., and M. Ryan. 1997. A physiological basis for biosphere-atmosphere interactions in the boreal forest: An overview. *Tree Physiology* 17:491–500.

Matson, P. A., P. M. Vitousek, J. J. Ewel, M. J. Mazzarino, and G. P. Robertson. 1987. Nitrogen transformations following tropical forest felling and burning on a volcanic soil. *Ecology* 68:491–502.

McGrath, D. A., M. L. Duryea, N. B. Comerford, and W. P. Cropper. 2000. Nitrogen and phosphorus cycling in an Amazonian agroforest eight years following forest conversion. *Ecological Applications* 10(6):1633–47.

McGrath, D. A., C. K. Smith, H. L. Gholz, and F. A. Oliveira. 2001. Effects of land-use change on soil nutrient dynamics in Amazônia. *Ecosystems* 4:625–45.

McNeil, M. 1966. Laterite soils. *Scientific American* 211:68–73.

Menaut, J. C., M. Lepage, and L. Abbadie. 1995. Savannas, woodlands and dry forests in Africa. In *Seasonally dry tropical forests.* Edited by S. H. Bullock. Cambridge, U.K.: Cambridge University Press, 64–92.

Neill, C., M. C. Piccolo, P. A. Steudler, J. M. Melillo, B. J. Feigl, and C. C. Cerri. 1999. Nitrogen dynamics in Amazon forest and pasture soils measured by ^{15}N dilution. *Soil Biology and Biochemistry* 31:567–72.

Nepstad, D. C., P. R. Moutinho, and D. Markewitz. 2001. The recovery of biomass, nutrient stocks, and deep soil functions in secondary forests. In *The biogeochemistry of the Amazon Basin.* Edited by M. E. McClain, R. L. Victoris, and J. E. Richey. New York: Oxford University Press.

Nobre, C. A., D. Wickland, and P. I. Kabat. 2001. The Large-Scale Biosphere-Atmosphere Experiment in Amazonia (LBA). *Global Change Newsletter* 45:2–4.

Nykvist, N. 2000. Tropical forest soils can suffer from a serous deficiency of calcium after logging. *Ambio* 29:310–13.

Perlin, J. 1989. *A forest journey: The role of wood in the development of civilization.* Cambridge, Mass.: Harvard University Press.

Reynolds, H. L., A. Packer, J. D. Bever, and K. Clay. 2003. Grassroots ecology: Plant-microbe-soil interactions as drivers of plant community structure and dynamics. *Ecology* 84:2281–91.

Richter, D., and L. I. Babbar. 1991. Soil diversity in the tropics. *Advances in Ecological Research* 21:315–89.

Richter, D. D., Jr., and D. Markewitz. 2001. *Understanding soil change: Soil sustainability over millennia, centuries, and decades.* Cambridge, U.K.: Cambridge University Press.

Sanchez, P. A., D. E. Bandy, J. H. Villachicia, and J. J. Nicholaides. 1982. Amazon Basin soils: Management for continuous crop production. *Science* 216:821–27.

Sanchez, P. A., and T. J. Logan. 1992. Myths and science about the chemistry and fertility of soils in the tropics. In *Myths and Science of Spoils of the Tropics.* Edited by R. Lal and P. Sanchez. Soil Science Society of America Special Publication 29. Madison, Wis.: Soil Society of America, 35–46.

Savage, K., T. R. Moore, and P. M. Crill. 1997. Methane and carbon dioxide exchanges between the atmosphere and the northern boreal forest soils. *Journal of Geophysical Research* 102:29, 279–29, 288.

Schlesinger, W. H. 1997. *Biogeochemistry: An analysis of global change.* San Diego: Academic Press.

Schroth, G., S. A. D'Angelo, W. G. Teixeira, D. Haag, and R. Lieberei. 2002. Conversion of secondary forest into agroforestry and monoculture plantations in Amazonia: Consequences for biomass, litter and soil carbon stocks after 7 years. *Forest Ecology and Management* 163:131–50.

Schroth, G., L. F. da Silva, R. Seixas, W. G. Teixeira, J. L. V. Macedo, and W. Zech. 1999. Subsoil accumulation of mineral nitrogen under polyculture and monocultural plantations and primary forest in a ferralitic Amazonian upland soil. *Agriculture, Ecosystems and Environment* 75:109–20.

Sellers, P. J., F. G. Hall, R. D. Kelly, A. Black, D. Baldocchi, J. Berry, M. Ryan, K. J. Ranson, P. M. Crill, D. P. Lettenmaier, H. Margolis, J. Cihlar, J. Newcomer, D. Fitzjarrald, P. G. Jarvis, S. T. Gower, D. Halliwell, D. Williams, B. Goodison, D. E. Wickland, and F. E. Guertin. 1997. BOREAS in 1997: Experiment overview, scientific results, and future directions. *Journal of Geophysical Research* 102:28, 731–28, 769.

Simard, S. W., D. A. Perry, M. D. Jones, D. D. Myrold, D. M. Durall, and R. Molina. 1997. Net transfer of carbon between ectomycorrhizal tree species in the field. *Nature* 388:579–82.

Sombroek, W. G. 1984. Soils of the Amazon Region. In *The Amazon: Limnology and landscape ecology of a mighty tropical river and its basin.* Edited by H. Sioli. Dordrecht, the Netherlands: D. W. Junk Publishers, 522–35.

Townsend, A. R., G. P. Asner, C. C. Cleveland, M. E. Lefer, and M. M. C. Bustamante. 2002. Unexpected changes in soil phosphorus dynamics along pasture chronosequences in the humid tropics. *Journal of Geophysical Research-Atmosphere* 107(D20):8067.

Trumbore, S. E., E. A. Davidson, P. Barbosa de Camargo, D. C. Nepstadt, and L. A. Martinelli. 1995. Below ground cycling of carbon in forests and pastures of eastern Amazonia. *Global Biogeochemical Cycles* 9:515–28.

Van Wambeke, A. 1992. *Soils of the tropics: Properties and appraisal.* New York: McGraw-Hill.

Vitousek, P. M., and H. Farrington. 1997. Nutrient limitation and soil development: Experimental test of biogeochemical theory. *Biogeochemistry* 37:63–75.

Wilson, E. O. 2002. *The future of life.* New York: Knopf.

SUGGESTED READINGS

Ewel, J., C. Berish, and B. Brown. 1981. Slash and burn impacts on a Costa Rican wet forest site. *Ecology* 62:816–29.

McGrath, D. A., M. L. Duryea, N. B. Comerford, and W. P. Cropper. 2000. Nitrogen and phosphorus cycling in an Amazonian agroforest eight years following forest conversion. *Ecological Applications* 10(6):1663–47.

McGrath, D. A., C. K. Smith, H. L. Gholz, and F. A. Oliveira. 2001. Effects of land-use change on soil nutrient dynamics in Amazônia. *Ecosystems* 4:625–45.

Sanchez, P. A., D. E. Bandy, J. H. Villachicia, and J. J. Nicholaides. 1982. Amazon Basin soils: Management for continuous crop production. *Science* 216:821–27.

HAPTER

4

A HUMAN GEOGRAPHY ERSPECTIVE ON TROPICAL)EFORESTATION

From Farmers to Satellites

Peter Klepeis

Over the past three decades, concern about tropical **deforestation** has become widespread, with images of bulldozers, burning trees, and expanding agriculture and pastureland entering the collective imagination. Cries of alarm abound in both the scholarly and popular press: organisms that may hold the key to curing cancer are being snuffed out before science can uncover their secrets, the carbon once held in the forest is being released into the atmosphere and is contributing to human-induced climate change, indigenous cultures are being transformed by exposure to the outside world.

Along with concerns about the possible negative impacts of deforestation on society, there are many assumptions about what is causing the problem in the first place. Conventional wisdom pins the blame on population growth and, in particular, poor farmers who colonize the forests and practice **shifting cultivation** (**slash-and-burn agriculture**). Other explanations focus on logging or ranching activities. But most perceptions of tropical deforestation do not involve careful review of the evidence, leaving the complexity of the change dynamics hidden.

Understanding environmental change requires answering some basic, although difficult, questions about pattern and process. How widespread is the phenomenon? What is the magnitude of the transformation? In the case of tropical forests, is clearing occurring around the globe at equal rates and with similar patterns? And what are the processes involved? Why are people cutting down trees? What mix of political, economic, cultural, and biophysical factors leads people to use the land in the way they do?

The discipline of geography has a long tradition of studying environmental issues and linking patterns of environmental change to their underlying processes. Geographers contrast characteristics that are common to tropical deforestation worldwide with those that are unique to particular places. And they consider societal response. What should the global community do about one of the most important environmental issues of our time?

Why Worry about Tropical Deforestation?

Tropical deforestation is a problem that receives serious attention by many segments of society—scientists, the media, policymakers, and students, among others. Numerous concerns are raised (Brown and Pearce 1994; Myers 1992; Place 2001). Tropical forests contain extremely high **biotic diversity**. As sources of wood, food, and drugs, they provide subsistence to people and maintain economies from local to global scales, and they serve as the homes of unique cultures. Tropical forests are also sources and **sinks** (or reservoirs) of carbon, which makes them an important part of global climate systems. But beyond being beneficial to society either directly, in the case of food production, or indirectly, by helping to maintain a climate conducive to humans, ethical considerations point to the intrinsic right of the forest to exist regardless of its utility to people.

With such a broad range of ecological, economic, cultural, and ethical considerations, it is no wonder that a corresponding broad range of researchers from the natural sciences, social sciences, and humanities study tropical deforestation. Any environmental change involves a web of multifaceted dynamics. The integration of multiple perspectives, sources of data, methodologies, and explanatory frameworks is necessary, therefore, if the complexity of deforestation processes is to be uncovered. Part of this diverse team of investigators includes geographers.

Research Foci in Geography

WHAT IS GEOGRAPHY?

It is common for people to think of geography as a discipline that catalogs names and other information about places in the world. Certainly, description of the earth's cultural and physical features is part of what geographers do. The environments of particular places consist of social, biological, and physical conditions, and geographers often display this information using the maps everyone associates with them. In addition to characterizing patterns, however, geographers seek to explain why such patterns exist.

Textbooks introduce geography as explaining the "why of where"; the assumption that location matters in natural and social processes underpins most analysis (National Research Council 1997). Why are tropical forests located where they are? What combination of factors produces forests in these specific locations? Why are they so rich in biotic diversity? Why is deforestation occurring where it is? What are the processes of environmental change affecting a given location?

Geographers are united by their efforts to connect landscape change with the multiple social and biophysical factors involved; in this way, geography represents, in a sense, a disciplinary middle ground. If geography were a three-legged creature, it would position itself with a foot in each of the broad categories of intellectual inquiry: the natural sciences, the social sciences, and the humanities.

Geography occupies a disciplinary middle ground in part because of its place-based focus. The **human–environment conditions** (the combination of social and biophysical characteristics) of a particular place are linked to overlapping factors (see figure 4.1). Sack (1997) argues that the ways in which people relate to their natural environment, and thus the ways they choose to manage natural resources, depend on forces in nature; social relations (the ways in which people interact with each other); and meaning, that which—apart from thoughts that are generated from our interaction with other people—is produced from within each of us.

This framework may seem abstract at first; however, think of how these forces shape the ways in which land management decisions are made. Natural forces and conditions place constraints on how humans behave. An area with more fertile soils and adequate precipitation facilitates agricultural production, whereas a dry zone with less-rich soils is a more challenging context for growing food. Clearly, aspects of how individuals live are also dependent on social relations, such as the rules and laws by which

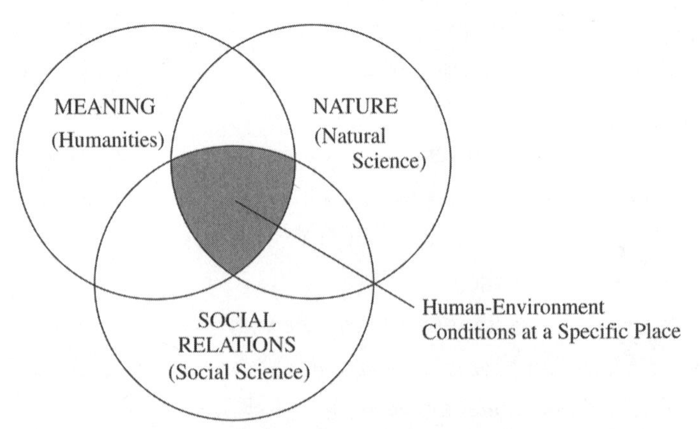

Figure 4.1. A central goal of research in geography is explaining environmental changes tied to specific places through interdisciplinary connections (after Sack 1997).

society is governed. Whose decision is it, for example, to cultivate a parcel of land or not—a government official, the head of household, or the collective community? Finally, the actions of individuals also depend on their worldviews, their notions of morality, and their sense of the aesthetic—the realm of meaning. Is there a spiritual connection to the forest for some people? Or do individuals hold the view that nature is there to serve the interests of human beings and is therefore to be exploited to the utmost? In short, geographers try to uncover how biophysical processes, social processes, and individual conceptualizations of nature intersect at a given place to cause a particular set of human-environment conditions.

PATTERNS

The world's tropical forests have undergone rapid transformation in recent decades, with the Food and Agriculture Organization (FAO) estimating an 8.7% (152 million hectare) reduction of tropical forest area worldwide between 1990 and 2000 (FAO 2001). Achard and colleagues (2002) estimate the global annual deforestation rate for humid tropical forests between 1990 and 1997 to be at 0.52%, although this rate drops to 0.43% if you consider net forest change—that is, both cutting and regrowth. In their analysis of 108 published studies of tropical deforestation worldwide, Geist and Lambin (2001, 76–78) put these figures in perspective by characterizing relatively low (0.3% to 0.7%) and relatively high (1.9% to 2.9%) mean annual deforestation rates by region. Rates do change through time. Kummer (1992) finds that deforestation rates in the Philippines between 1947 and 1987 range from 1.74% per year to upwards of 3.64% per year, with an overall loss of forest estimated at approximately 84,900 square kilometers. Cortina and colleagues (1999) find that between 1975 and 1981, the annual rate of deforestation in the southern Yucatán Peninsula was an extremely high 5.2% per year but dropped to 1.4 % per year between the mid-1980s and 1990. The average rate of deforestation between 1969 and 1997 for the same general region, however, was a relatively low 0.4% per year (Turner et al. 2001). But while improvements continue to be made, assessments of forest cover change remain crude, flawed in part by inadequate monitoring, inconsistent definitions of what constitutes deforestation, and poor data quality (Fairhead and Leach 1998; Matthews 2001). There is consensus that tropical deforestation is a serious problem and that the scale of human impact is great; however, our understanding of where and when forest disturbance has occurred is incomplete.

There are at least four reasons why determining deforestation rates across time and space is challenging. First, the extent of tropical forests worldwide is large, covering some 1.74 billion hectares (FAO 2001). It is difficult to know what kinds of impacts are occurring across such a vast expanse, in particular given the geographical isolation of many forests. Ground-level surveys are helpful—if carried out effectively—but they only provide documentation on human impact for select locations. Second, the cost of mounting comprehensive ground surveys may be prohibitive, making continuous monitoring difficult, especially given that countries with tropical forests are often poor. Third, there is debate over how to define what is meant by tropical deforestation. If a specific unit area of forest has 90% of its canopy cleared (the FAO's definition), is that area considered "deforested"? Is an area that is cleared of forest one year but that experiences regrowth in subsequent years still defined as deforested? What about other kinds of human impacts on forests that don't involve clear-cutting, such as selective logging or partial cuts? It is difficult to compare estimates of forest disturbance, especially those by different researchers, without consistent answers to these questions. Finally, inadequate or unknown baseline data about the condition of forests long ago, before significant human occupation of them, limits what researchers can say about overall human impact on forests.

The analysis of **remotely sensed imagery** (information from satellites or aerial photographs) using **geographic information systems** (GIS) addresses some of these challenges and enhances the assessment and monitoring of deforestation patterns. A GIS is a computer-based technology that allows researchers to analyze, manage, and display spatial information, such as satellite imagery or road networks. Imagine a stack of paper maps each representing a different theme: political boundaries; soil characteristics; population density; precipitation patterns; **land cover** (e.g., cleared land, forest, grassland, wetland); and other phenomena that vary across space. A GIS allows the researcher to manage digital versions of these data layers and to analyze the importance of each for a particular environmental issue.

The use of satellite imagery and GIS for monitoring deforestation allows the vast expanse of tropical forests to be monitored much more cheaply in comparison to ground-level surveys. And by analyzing imagery from multiple periods over the past three decades (satellite imagery became available for research in the mid-1970s), investigators can also identify the rate of deforestation (area cleared per year) and patterns of forest disturbance through time. In addition, GIS and remote sensing help uncover the dynamics of deforestation. A simple and intuitive example is the finding that where there are roads and human settlements, there tends to be more forest disturbance than in isolated and thinly populated areas. But less obvious connections become apparent as well; **topography** (slope, aspect, elevation) or farm size, for example—variables that may influence the viability of agriculture—are often important in determining whether settlers choose to clear and cultivate a given location.

PROCESSES

GIS and remote sensing, by their very nature, involve analysis of spatially explicit and quantifiable data that are mapped easily onto specific locations. But there are other factors driving deforestation that are qualitative and not spatially explicit—that is, they cannot be tied to specific geographic coordinates. To fully answer why and how tropical deforestation occurs, therefore, requires additional approaches, such as surveying local land managers to understand their decision-making processes, exposing the history of development policy through archival analysis, and directly observing regional land use systems.

There is significant debate over the degree to which variables at the local or community scale (e.g., population growth) drive most of the deforestation, versus factors that manifest themselves at regional, national, or international scales, such as government policies or demand for commodities produced in the tropics (e.g., coffee, timber, rubber). Put another way, mass media often blame tropical deforestation on farmers who use shifting cultivation to feed their families, while **nongovernmental organizations** (NGOs) and environmental activists tend to blame government policies or logging companies. Who is right?

It is clear from this question that knowing the why, where, and how of tropical deforestation is more than an interesting environmental problem. The answer influences how society chooses to respond. If shifting cultivation is the main culprit, should alternative livelihood opportunities be made available to forest inhabitants? Can forest cover and biotic diversity be maintained while also providing income and subsistence for local people?

The kinds of answers to the "what is causing tropical deforestation?" question, minimally, can be split into two broad kinds (Klepeis 2003). The first focuses on **human agency**, or the ability of individuals and small groups to manipulate conditions to their own benefit. In this form of explanation, the immediate agents of change (i.e., local-scale land managers and inhabitants) are identified as the primary players driving deforestation. The assumption is that regional population growth puts increasing pressure on human-environment conditions, leading inhabitants to open up frontier forests to expand their livelihood opportunities. A second form of explanation focuses on **social structures**, or the rules, conventions, and restraints that govern individual human behavior. In this case, political and economic conditions (defined by a mix of social structures) are shown to empower certain economic agents (e.g., logging or mining companies) that clear the forest with little regard for the social and ecological consequences, but at the same time to disempower the forest inhabitants and their capacity to manage the forest sustainably. Increasingly, it is recognized that both of these forms of explanation—structure and agency—reflect the human-environment conditions causing deforestation around the world.

Studies of the role of human agency in deforestation tend to focus on local- or proximate-scale dynamics. Throughout history, time and again, people have shown themselves remarkably adaptive to shifting human-environment conditions. Part of adaptation involves transforming the environment in ways that suit societal needs. For example, from Australia to Africa to the Americas, and in a broad range of biophysical contexts, indigenous groups have used fire to create advantageous conditions for hunting and agriculture. Whatever the mix of flora and fauna and natural resources available in a particular location, humans have shown themselves capable of exploiting the environment to their advantage. In tropical forests, an example of this effective adaptation is shifting cultivation, a system that is well suited to the ecology of the forest and that represents an efficient mode of food production (see below). But in the context of human-environment conditions that change too fast, such as the 20th-century boom in population growth and frontier settlement, the capacity for adaptation is weakened: short-term livelihood needs can displace long-term, more sustainable land management strategies. And, if no alternative livelihood opportunities are available, the result usually includes clearing more forest.

Analysis of social structures shifts the focus away from just the local scale and considers how regional inhabitants may be forced to clear forests due to factors beyond their control, emanating from outside their sphere of influence (Folke et al. 1998). Two important examples are the influence of government policies and markets. Government policies affect individual behavior by making some activities illegal and by encouraging others deemed advantageous to society, and by segregating access to resources for different groups of people (e.g., taking away indigenous land and leasing it to logging companies or setting it aside as a nature reserve). Federal policies that encourage forest clearing through tax incentives; that provide roads allowing access to frontier locations; or that invest in particular land uses, such as large-scale ranching initiatives, take agency away from local inhabitants. And fluctuations in market demand for particular forest commodities, such as rubber, can negatively affect local livelihoods, forcing inhabitants to search for alternative land uses. In essence, explaining deforestation in terms of social structures asks you to consider local-scale dynamics as more than a closed system, and to follow the chain of linkages away from where forest clearing occurs to the national and international markets and the federal development policies that underlie the behavior of local land managers.

In reference to the widely assumed link between population growth and environmental degradation, for example, analysis of social structures shows that population growth in an area undergoing deforestation is often due to social dynamics elsewhere. In many cases, an influx of people from outside the region, rather than natural population growth (as births exceed deaths), explains the population change. Land scarcity or lack of employment opportunities in urban areas may cause the federal government to use colonization programs in forest frontiers as a release valve to diffuse the political tension created by widespread poverty and unemployment. The Brazilian government moved people from coastal regions into the Amazon basin in the 1970s and 1980s to provide opportunities for disenfranchised people as well as to intensify the exploitation of the country's natural resources in the hopes of fostering economic development (Hecht 1985).

Current Research in the Field

Research on the patterns and processes of tropical deforestation has led to important insights that hold implications for possible societal responses to the problem. From this large body of work, four themes are emphasized: 1) perceptions of shifting cultivation, 2) forest regeneration or secondary growth, 3) boom-bust cycles of deforestation, and 4) strategies to combine forest conservation with economic development. These themes are representative of research in multiple contexts across the world, although much of the evidence referred to here is from Latin America. A case from southeastern Mexico is used here to exemplify the crosscutting themes in deforestation research and efforts to integrate various forms of explanation in a single research project (see box 4.1).

Box 4.1. The SYPR Project: Linking Pattern to Process

The land cover and land use change—The Southern Yucatán Peninsular Region (SYPR) project is a collaborative research venture between Clark University, Harvard Forest (Harvard University), and El Colegio de la Frontera Sur, a research institute in Mexico. The 22,000-square-kilometer study region is in southeastern Mexico, abutting the borders of Guatemala and Belize. It is part of the largest contiguous tropical forest remaining in Mexico and Central America. The region underwent widespread deforestation in the distant past (by the Classic Maya civilization, AD 300–900), and, since the 1960s, there have been high rates of deforestation due to extensive road building, colonization and land grants, large agricultural projects, and expanding small-scale agriculture (Turner et al. 2004). In seeking to explain both the patterns and the processes of deforestation, the SYPR project links two research approaches: in-depth ground-level studies that focus on case studies, detailed assessments of human-environment conditions, and social and biophysical change processes; and the use of satellite imagery and aggregate socioeconomic data to capture broad patterns of forest disturbance.

Patterns of deforestation—Between 1969 and 1997, the total **anthropogenic disturbance** (that caused by humans) approached 10%. The region underwent a boom in forest clearing in the 1970s and early 1980s due to government-directed colonization and large agriculture projects for rice and cattle in lowland areas, as well as government-subsidized clearing for shifting cultivation in upland forests. There was a lull in forest clearing after these large projects failed and government investment dried up, only to be followed by a subsequent increase in deforestation rates in the face of spontaneous colonization by settlers from land-scarce regions in other parts of Mexico. But, over the past 10 years, now that most land in the area has been allocated, rates have again slowed. Much of the forest that was once cleared has reverted to multiple stages of secondary growth. For the 1995–2001 period, satellite imagery shows less land in cultivation than in succession or fallow and a reduction in the cutting of mature forest.

Processes of deforestation—20th-century trends in deforestation are primarily linked to structural shocks and, until recently, not the agency of local farmers. Government policies opened up the region to international forest companies

(*continued*)

Box 4.1. (*Continued*)

in the first half of the 20th century, leading to the construction of road networks that enabled subsequent settlement. In the 1970s and 1980s, policies targeted the area for settlement and agricultural development. One of the reasons for the focus on agriculture and not on the selective logging of mahogany and the extraction of chicle (a tree resin that serves as a base for chewing gum) was the prior depletion of mahogany stands due to shortsighted management and the bust of the chicle market after a synthetic chewing gum base was created after World War II (activities surrounding chicle exploitation led to minimal forest disturbance). Following the bust in forestry, the failure of the large-scale agriculture and ranching initiatives, and the decline in land available for new settlers, deforestation rates slowed.

Perceptions of human impact on the forest—The species composition of mature forests today bears the mark of Classic Maya activity from over 1,000 years ago and from selective logging from the mid-20th century. Recognition of the human element in the forest's history may partly explain why most conservation initiatives in the region now recognize the need for an integrated land management strategy that embraces both sustainable economic uses of the forest and nature reserves.

National policy responses—Government attempts to slow deforestation are mixed at best but include the creation of the Calakmul Biosphere Reserve in 1989, incentives to encourage more intensive agriculture (to maintain or raise food production without clearing new land), and investment in an ecological and archeological tourism scheme. The more likely factors in the lower rates, however, are the lack of federal investment in new projects that involve major clearing; the stabilization in regional population; farmer-led experiments with new agricultural systems (e.g., jalapeño chili production); and the rise of community forestry projects.

Note: Major funding for the SYPR project (1997–2000) was from NASA's LCLUC program and the Center for the Integrated Studies of the Human Dimensions of Global Change (Carnegie Mellon University, NSF).

PERCEPTIONS AND ADAPTATION

Three important findings about tropical forests and their inhabitants are calling into question some long-standing assumptions: 1) traditional food production systems, such as shifting cultivation, *are* often well adapted to human-environment conditions; 2) assumptions that tropical forests have an intrinsically low **carrying capacity**—a term that notes limits on the capacity of the environment to support a growing population—ignore the many historical examples of intensive agriculture in tropical forest locations where high population densities were supported; and 3) tropical forests are *not* "virgin" or "pristine" but instead contain a human imprint both from ancient and more recent activities.

First, despite conventional wisdom that shifting cultivation is a maladaptive system responsible for degrading forests, this form of agriculture is energy and labor efficient and is well adapted to a tropical forest ecology, mimicking the forest's complex structure and high biotic diversity (Denevan 2001, 83–90; Rappaport 1971). Shifting cultivation involves cutting the vegetation for a given plot of land, allowing it to dry, and burning it to both clear the land of vegetation and to enrich the soil with ash. Traditionally, after a few years, as soil fertility declines and weeds become a problem, the land is allowed to revert to forest over the course of many years, a phase in which the land is in fallow. In the meantime, the farmer repeats the process in numerous other locations and then returns to the original spot after 10 to 20 years. In areas with low population density, where forested land is plentiful, the system can sustain production levels through time without disrupting the forest's basic structure and composition. Indeed, Rappaport (1971) finds a lesson in shifting cultivation for advancing sustainable agriculture today, in that it maintains diverse, complex agroecosystems that are more resistant to pests than are the monocropped fields of modern agriculture. There is a cautionary tale, however. As population densities increase and land becomes scarcer, the length of the forest fallow period decreases, which may lower biotic diversity, crop yields, and the forest's ability to recover.

This potential for declining agricultural productivity in the face of increasing population and land pressure has led many researchers to believe that tropical forests have a low carrying capacity. The argument is that most tropical environments are not suited for intensive occupation and land use (e.g., because of pests, thin soils, and high humidity that make food storage difficult), and thus human population levels must remain small. But carrying capacity is determined largely by technologies and land management strategies rather than by environmental conditions (Brush 1975).

The southern Yucatán Peninsula in southeastern Mexico provides a case in point (Turner 1990). Current population densities in the region are quite low, averaging only a few people per square kilometer. In contrast, average population densities during the Classic Maya period (AD 300–900) approached 100 people per square kilometer, with those in urban areas exceeding 1,000 people per square kilometer. To maintain such high population densities, the Classic Maya civilization—which was spread throughout the tropical lowlands of southern Mexico, Guatemala, Belize, Honduras, Nicaragua, and El Salvador—used intensive and complex agricultural systems. It did not rely solely on shifting cultivation. The lesson is that, while environmental conditions in tropical forest locations may place constraints on agricultural production, technical sophistication and adaptive resource management can raise the capacity for food production. The concept of carrying capacity is, therefore, best used with caution; it is helpful to the degree that it suggests limits to the natural resource base (e.g., topsoil, groundwater, biotic diversity), but it should not be used to indicate some absolute number of people that nature can support.

Finally, there is a pervasive assumption that tropical forests were sparsely occupied before the 20th century and that significant human impact on the forests is a recent phenomenon. Denevan (1992) calls this assumption the **pristine myth**. But, in contrast to notions that precolonial forests were untouched by people and were pristine, humans have a long history of transforming tropical forest landscapes in many parts of the world (Denevan 2001; Whitmore and Turner 2001). The legacy of this impact is seen in forest structure and species composition; in archeological evidence of intensive agricultural systems (e.g., terraces and raised fields); and in the evidence of anthropogenic soils (deep, rich, fertile soils created by people practicing intensive agriculture) (Denevan 2001). For example, one estimate is that *terra preta*—"black earth," a very fertile soil created by indigenous groups long ago—covers 10% of Amazonia (Mann 2002).

Returning to the southern Yucatán Peninsula example, the Classic Maya civilization had almost completely deforested the region by AD 900–1000, after which it collapsed due to a combination of economic, political, and environmental factors (Turner et al. 2003). Subsequently, human impact on the land became miniscule, and mature tropical forests returned over the course of hundreds of years. In other words, the forest we see today is essentially secondary growth. And its structure and species composition bear the human imprint of the Classic Maya (e.g., Maya orchard gardens built up economically useful species that are still found today).

Knowing forest histories, such as that of the Classic Maya region, is important. Our assumptions about whether or not the forest is pristine affects how we choose to manage it today. If places once thought to have existed for millennia as primordial forest are found to be human creations, at least in part, then arguments to preserve them, away from people, become problematic. If the very nature that we celebrate as "wilderness" carries a human imprint, then why not see people as part of nature? This is not to say that we should not worry about deforestation and should permit widespread clearing; however, it does suggest that we should be able to figure out a way for people to coexist with forests (see section on "societal response" below).

FOREST REGROWTH AND FISH BONES

"Forests are not static, but living entities; they regenerate and grow in favorable environmental conditions. They can be replanted as well as removed, managed as well as neglected, and their productivity hindered or encouraged" (Williams 1990, 195).

The degree of human impact on forests varies across space and time. The spatial pattern of forest disturbance captured in remotely sensed imagery provides insights into processes of land use change, such as the role of directed colonization schemes. But imagery most effectively detects forested and nonforested areas, simple categories that hide the complexity of heterogeneous landscapes. Human use and natural perturbations disturb forest and lead to multiple stages of succession, from very young to mature regrowth. These patches of secondary growth are often the product of shifting cultivation, which lets forest shift in and out of use over time; land abandonment; or biophysical shocks such as fires and hurricanes. Two other land use systems that make for a diverse landscape are **agroforestry**, a form of land management that incorporates a mix of trees and crops, and **forest plantations**, which are cultivated forests that lack species diversity but that represent areas of maintained forest cover nonetheless. Explanations of the patterns of deforestation should reflect the dynamic nature of forests and their use.

The complexity of land uses is not always easy to detect in a satellite image, which is why researchers are often forced to rely on forest-nonforest designations. Despite this limitation, the patterns of forest clearing captured by remotely sensed imagery suggest underlying processes of change. A widely recognized symbol of tropical deforestation is the fish-bone pattern (see figure 4.2). While explanations of the pattern are fairly simple, the example underscores the importance of integrating analysis of remotely sensed imagery with intensive ground-level studies—the practice of linking pattern to process. A common scenario has the federal govern-

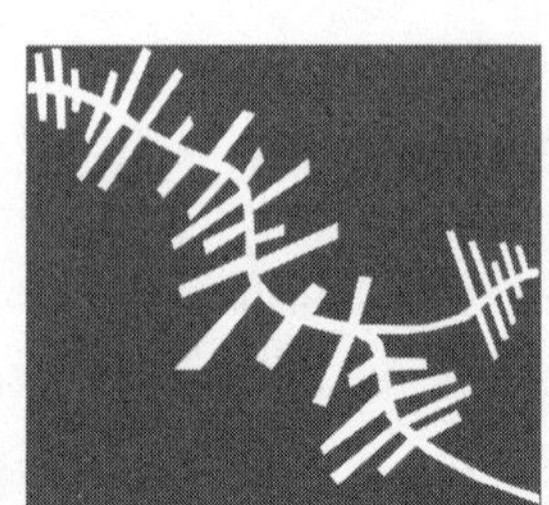
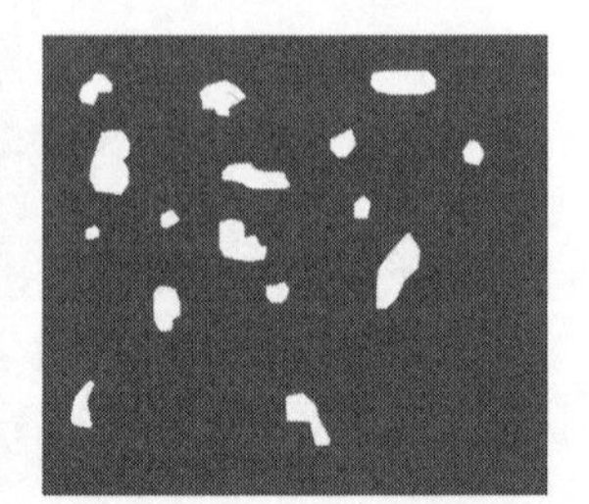

Figure 4.2. Three forest-nonforest spatial patterns (after Geist and Lambin 2001: 66).

ment of a developing country encouraging the settlement of frontier forests to exploit land resources and to foster economic development. A road is built that connects urban centers with the region to be settled. The government explains to settlers that they will receive title to those lands that are cleared of trees because it seeks to promote agricultural production for the market. Settlers arrive via the road and gain access to parcels of land with rectangular shapes. People tend to clear trees beginning close to the road and then gradually proceed away from it. This scenario explains why remotely sensed imagery shows rectilinear clearings adjacent to the road. It also shows how explaining the patterns of deforestation seen in imagery requires understanding the underlying factors in forest clearing, such as government policy. Two other examples of patterns found in imagery are geometric patterns, which reflect large-scale clearing activities such as cattle ranching, and diffuse patterns, which reflect smallholder agricultural and usually shifting cultivation.

The study of patterns of deforestation and the phenomenon of forest regeneration are closely linked. For example, forest recovery (regrowth) in temperate areas of the world tends to occur on peripheral lands, away from roads. This occurs in most cases because land that was once used for agriculture is abandoned due to increased off-farm labor opportunities in more urban areas. Rudel, Bates, and Machinguiashi (2002) find, however, that in the Ecuadorian Amazon, regrowth is occurring close to roads. Farmers with small landholdings are abandoning cattle ranching, which often occurred close to the road, to practice shifting cultivation. The reason is that markets have changed such that agricultural produce for urban markets is in greater demand than livestock. The farmers have an incentive, therefore, to use their lands for agriculture as opposed to ranching, which necessitates changing the land use system and allowing much of the land near the road to revert to forest fallow (unless chemical fertilizers and pesticides are readily available, in which case the land near the road may be used for agriculture right away). Presumably, this secondary growth will be used for shifting cultivation as times goes by.

Despite assumptions that deforestation inevitably leads to devastation, such as permanent conversion from forest to pasture, in some areas of the tropics, rates of forest recovery are becoming higher than the rates of ongoing forest loss, resulting in a net gain of forest cover (Perz and Skole 2003). It is true that secondary forests are not the same as mature ones, but they are not ecologically trivial. Under the right human-environment conditions and in as little as three decades, biomass, canopy height, and even species composition can start to approach precut levels (Finegan and Nasi 2004). The recovery of sections of once deforested areas underscores how forest change is not a linear process. Moran and Brondizio (1998) find that many cut or burned areas experience rapid regrowth and that these patterns of regrowth can be linked to differences in soil fertility. Skole and colleagues (1994, 316) show that 42% of new agricultural land created between 1988 and 1989 in the Brazilian state of Rondônia came from clearing secondary growth. Recent work on deforestation in the southern Yucatán Peninsula (see box 4.1) reinforces the importance of understanding how farmers and other land managers use secondary forest. Significant tracts of mature forest were cut in the 1970s and 1980s; however, in the 1990s, farmers tended to clear successional growth (Turner et al. 2001). Research into why this is so holds implications for how more sustainable uses of the forest might include use of secondary forest as opposed to more mature vegetation.

These examples demonstrate the variability in the extent of deforestation and the need to specify carefully the spatial and temporal scale of analysis when discussing human impacts on forests. Rates fluctuate across space and time due to variability in the forces driving the deforestation. Shifts in the demand for forest products, such as in the Ecuadorian case; changes in government policy (e.g., the implementation or cancellation of a colonization program); or increased pressure to conserve **biodiversity** may factor into whether there is more or less forest clearing in any given historical period.

BOOM-BUST CYCLES

Fluctuating rates of deforestation through time reflect shock and intershock periods. Shocks often manifest themselves as sudden shifts in government policy and commodity prices (Kummer and Turner 1994). The classic example of a shock is the construction of a new road into a previously unoccupied or thinly occupied area. Almost overnight, an inaccessible forest is opened up to newcomers, whether they be loggers, miners, ranchers, or subsistence farmers. Likewise, a shift in market demand for beef, timber, grain, minerals, or some other resource that might be extracted from the land may lead to new interest in opening forested regions for economic use. But shocks do not always lead to increased forest disturbance. The rise in international conservationism in the 1980s and 1990s led many governments to establish nature reserves. The effectiveness of these conservation strategies is mixed; however, in many instances, reserves led to reduced deforestation rates for select locations.

Deforestation is also tied to forces operating between the shock periods, such as factors involving the agency of local people. Coomes and Barham (1997) show that forest inhabitants in Peru frequently change their livelihood strategies over the course of their lifetime depending on access to land, labor, and capital. Land management is also explained by analyzing intrahousehold dynamics, such as differences in the roles of men and women in managing forest resources, or the age structure of household occupants. Sadoulet, de Janvry, and Davis (2001) find that the degree to which Mexican farmers intensify agricultural production (and reduce reliance on extensive shifting cultivation) depends largely on their access to capital with which to invest in production, to extension services that introduce new land management strategies, and to local organizations that allow farmers to make linkages to markets outside the region.

SOCIETAL RESPONSE

Linking pattern to process can be contentious, as demonstrated in a recent debate in the journal *Science* about the causes of deforestation in Amazonia. One group of researchers (Laurance et al. 2001, 2005) identifies highways and roads as the primary factor driving deforestation. While acknowledging the importance of road infrastructure, other researchers (e.g., Bruna and Kainer 2005; Schaeffer and Rodrigues 2005) express concern over the implication of what they see as an overly simplistic finding: slow road creation and you'll slow deforestation rates. They call for a recognition of both the broad suite of underlying forces of Amazonian development, and therefore the acknowledgment that multifaceted societal responses are necessary to address the deforestation problem, and the role that market access (via roads) has in achieving a balance between environment and development in the region.

In considering the possibility of balancing environment and development goals, scientists, politicians, and the general public invariably rely on the concept of **sustainable development**, the notion that economic growth and material development can occur at the same time as social welfare, equity, and environmental conditions are maintained or improved (Brundtland 1987). But is sustainable use of tropical forest resources possible? The assumption of those who promote sustainable development is that local people must be able to earn a good living from tropical forests if they are to have the incentive to stop clearing it. The use of **extractive reserves** is one popular idea to this end.

Extractive reserves are tracts of forest in which forest clearing is prohibited. Local people are allowed to enter the reserve to extract forest products, such as rubber, chicle, or nuts, or to selectively log. These forest commodities are then sold in regional, national, or international markets to support local livelihoods and to offer alternatives to uses that require clearing the forest.

In the short term, if demand for extractive forest commodities is high, extractive reserves work fairly well. Analysis over the long term paints a different picture, however. Markets in forest commodities are not stable, fluctuating dramatically over time (Coomes 1995). In the cases of chicle and rubber, synthetic substitutes were created that caused demand for these forest products to plummet. What are locals to do if their source of income disappears? They search for other strategies to make ends meet, such as agriculture, ranching, or off-farm labor.

But if extractive reserves cannot prevent deforestation from occurring, then what should be society's response? The answer you get depends on the person you ask. Researchers in geography and conservation biology increasingly call for diversified land management systems that integrate the use of nature reserves with sustainable economic uses. In other words, protected areas must be tied to poverty alleviation. Some areas are so ecologically sensitive that biotic diversity and other **ecological services** cannot be subject to human use without degradation occurring. But understanding of how to use the land sustainably is improving. Integrating extractive industries with intensive agriculture, agroforestry, and other productive land uses, along with the provision of livelihood strategies not necessarily connected to the land, decreases the vulnerability of local people to poverty and reduces the prospect that they will be forced to cut the forest unsustainably (Browder 2001). If diversified, the capacity of the regional economy to absorb fluctuations in commodity prices or other perturbations is enhanced, which is good for both the people and the environment. How to provide this diversity is subject to considerable debate. However, researchers increasingly recognize the need for democratic decision making; "active, inclusive, and iterative communication" between regional stakeholders tends to lead to more informed decisions, more sustainable use systems, and a better balance between environment and development (Cash et al. 2003, 8088).

Conclusion

Research in geography shows clearly that understanding tropical deforestation dynamics necessitates making local to global, cross-scalar connections. Deforestation is caused by a web of socioeconomic, political, and biophysical factors that manifest themselves at intrahousehold, parcel, community, regional, national, and international scales. But by comparing case studies and context-specific deforestation dynamics with global patterns and trends, geographers are starting to navigate their way through this complexity and to improve our understanding of human-environment relationships. By exposing the range of factors that affect local production systems, geographers are participating in a larger discourse on how to achieve sustainable development. By debunking the myths of simplistic, linear causal explanatory chains, research findings underscore the lack of a silver-bullet solution to the deforestation problem. What is left is the challenging process of designing diverse landscapes that simultaneously fit human and ecosystem needs.

REFERENCES

Achard, F., H. Eva, H. J. Stibig, P. Mayaux, J. Gallego, T. Richards, and J. P. Malingreau. 2002. Determination of deforestation rates of the world's humid tropical forests. *Science* 297:999–1002.

Browder, J. O. 2001. Alternative rainforest uses. In *Tropical rainforests: Latin American nature and society in transition.* Edited by S. Place. Jaguar Books on Latin America, no. 2. Wilmington, Del.: Scholarly Resources, 196–205.

Brown, K., and D. W. Pearce., eds. 1994. *The causes of tropical deforestation: The economic and statistical analysis of factors giving rise to the loss of the tropical forests.* London: University College London.

Bruna, E. M., and K. A. Kainer. 2005. A delicate balance in Amazonia. *Science* 307:1044–55.

Brundtland, G. H. 1987. *Our common future: World Commission on Environment and Development.* Oxford: Oxford University Press.

Brush, S. B. 1975. The concept of carrying capacity for systems of shifting cultivation. *American Anthropologist* 77:799–811.

Cash, D. W., W. C. Clark, F. Alcock, N. M. Dickson, N. Eckley, D. H. Guston, J. Jager, and R. B. Mitchell. 2003. Knowledge systems for sustainable development. *Proceedings of the National Academy of Sciences* 100(14):8086–8091.

Coomes, O. T. 1995. A century of rain forest use in western Amazonia. *Forest Conservation and History* 39(3):108–20.

Coomes, O. T., and B. L. Barham. 1997. Rain forest extraction and conservation in Amazonia. *Geographical Journal* 163(2):180–88.

Cortina, V. S., P. Macario Mendoza, and Y. Ogneva-Himmelberger. 1999. Cambios en el uso del suelo y deforestacion en el sur de los estados de Campeche y Quintana Roo, Mexico. *Boletin del Instituto de Geografia de la UNAM* 38:41–56.

Denevan, W. M. 2001. *Cultivated landscapes of native Amazonia and the Andes.* Oxford: Oxford University Press.

———. 1992. The pristine myth: The landscapes of the Americas in 1492. *Annals of the Association of American Geographers* 82:369–85.

Fairhead, J., and M. Leach. 1998. *Reframing deforestation: Global analyses and local realities.* New York: Routledge.

FAO (Food and Agriculture Organization of the United Nations). 2001. *Global forest resources assessment 2000.* FAO forestry paper 140. Rome: FAO.

Finegan, B., and R. Nasi. 2004. The biodiversity and conservation potential of shifting cultivation landscapes. In *Agroforestry and biodiversity conservation in tropical landscapes.* Edited by G. Schroth, G. A. B. da Fonseca, C. A. Harvey, C. Gascon, H. L. Vasconcelos, and A. N. Izac. Washington, D.C.: Island Press.

Folke, C., L. Pritchard Jr., F. Berkes, J. Colding, and U. Svedin. 1998. *The problem of fit between ecosystems and institutions.* IHDP working paper, no. 2. Bonn, Germany: International Human Dimensions Programme on Global Environmental Change (IHDP).

Geist, H. J., and E. F. Lambin. 2001. *What drives tropical deforestation? A meta-analysis of proximate and underlying causes of deforestation based on subnational case study evidence.* LUCC Report Series, no. 4. Louvain-la-Neuve, Belgium: LUCC International Project Office.

Hecht, S. B. 1985. Environment, development and politics: Capital accumulation and the livestock sector in eastern Amazonia. *World Development* 13(6):663–84.

Klepeis, P. 2003. Development policies and tropical deforestation in the southern Yucatán Peninsula: Centralized and decentralized approaches. *Land Degradation and Development* 14:1–21.

Kummer, D. M. 1992. *Deforestation in the postwar Philippines.* University of Chicago geography research paper, no. 234. Chicago: University of Chicago Press.

Kummer, D. M., and B. L. Turner II. 1994. The human causes of deforestation in Southeast Asia. *BioScience* 44(5):323–28.

Laurance, W. F., M. A. Cochrane, S. Bergen, P. M. Fearnside, P. Delamonica, C. Barber, S. D'Angelo, and T. Fernandes. 2001. The future of the Brazilian Amazon: Development trends and deforestation. *Science* 291:438–39.

Laurance, W. F., P. M. Fearnside, A. K. M. Albernaz, H. L. Vasconcelos, and L. V. Ferreira. 2005. Amazon deforestation models—response. *Science* 307:1044.

Mann, Charles C. 2002. 1491. *Atlantic Monthly*, March, 41–53.

Matthews, E. 2001. *Understanding the Forest Resources Assessment 2000.* Forest Briefing, no. 1. Washington, D.C.: World Resources Institute.

Moran, E. F., and E. Brondizio. 1998. Land-use change after deforestation in Amazonia. In *People and pixels: Linking remote sensing and social science.* Edited by D. Liverman, E. F. Moran, R. R. Rindfuss, and P. C. Stern. National Research Council. Washington, D.C.: National Academy Press, 94–120.

Myers, N. 1992. *The primary source: Tropical forests and our future.* New York: Norton.

National Research Council. 1997. *Rediscovering geography: New relevance for science and society.* Washington, D.C.: National Academy Press.

Perz, S. G., and D. L. Skole. 2003. Secondary forest expansion in the Brazilian Amazon and the refinement of forest transition theory. *Society and Natural Resources* 16:277–94.

Place, S. E., ed. 2001. *Tropical rainforests: Latin American nature and society in transition.* 2nd ed. Wilmington, Del.: Scholarly Resources.

Rappaport, R. 1971. The flow of energy in an agricultural society. *Scientific American* 225:116–32.

Rudel, T. K., D. Bates, and R. Machinguiashi. 2002. A tropical forest transition? Agricultural change, out-migration, and secondary forests in the Ecuadorian Amazon. *Annals of the Association of American Geographers* 92(1):87–102.

Sack, R. D. 1997. *Homo geographicus: A framework for action, awareness, and moral concern.* Baltimore, Md.: Johns Hopkins University Press.

Sadoulet, E., A. de Janvry, and B. Davis. 2001. Cash transfer programs with income multipliers: PROCAMPO in Mexico. *World Development* 29(6):1043–56.

Schaeffer, R., and R. L. V. Rodrigues. 2005. Underlying causes of deforestation. *Science* 307:1046.

Skole, D., W. H. Chomentowski, W. A. Salas, and A. D. Nobre. 1994. Physical and human dimensions of deforestation in Amazonia. *BioScience* 44(5):314–22.

Turner, B. L., II. 1990. The rise and fall of Maya population and agriculture, 1000 B.C. to present: The Malthusian perspective reconsidered. In *Hunger and history: Food shortages, poverty and deprivation.* Edited by L. Newman. Oxford: Basil Blackwell, 178–211.

Turner, B. L., II, J. Geoghegan, and D. R. Foster, eds. 2004. *Integrated land-change science and tropical deforestation in the southern Yucatán: Final frontiers.* Oxford: Oxford University Press.

Turner, B. L., II, P. Klepeis, and L. S. Schneider. 2003. Three millennia in the southern Yucatán peninsular region: Implications for occupancy, use, and "carrying capacity." In *Lowland Maya area: Three millennia at the human-wildland interface.* Edited by A. Gómez-Pompa, M. Allen, S. Fedick, and J. Jimenez-Osornio. New York: Haworth Press, 361–87.

Turner, B. L., II, S. C. Villar, D. Foster, J. Geoghegan, E. Keys, P. Klepeis, D. Lawrence, P. M. Mendoza, S. Manson, Y. Ogneva-Himmelberger, A. B. Plotkin, D. P. Salicrup, R. R. Chowdhury, B. Savitsky, L. Schneider, B. Schmook, and C. Vance. 2001. Deforestation in the southern Yucatán peninsular region: An integrative approach. *Forest Ecology and Management* 154(3):353–70.

Whitmore, T. M., and B. L. Turner II. 2001. *Cultivated landscapes of Middle America on the eve of conquest.* Oxford: Oxford University Press.

Williams, M. 1990. Forests. In *The earth as transformed by human action.* Edited by B. L. Turner II, W. Clark, R. Kates, J. Richards, and W. Meyer. Cambridge: Cambridge University Press, 179–201.

SUGGESTED READINGS

Bray, D. B., and P. Klepeis. Forthcoming. Deforestation, forest transitions, and institutions for sustainability in southeastern Mexico, 1900–2000. *Environment and History.*

Lambin, E. F., and H. J. Geist. 2003. Regional differences in tropical deforestation. *Environment* 45(6):22–36.

Mann, Charles C. 2002. 1491. *Atlantic Monthly,* March, 41–53.

O'Brien, K. L. 1998. *Sacrificing the forest: Environmental and social struggles in Chiapas.* Boulder, Colo.: Westview Press.

Place, Susan E. 2001. *Tropical rainforests: Latin American nature and society in transition.* Jaguar Books on Latin America, no. 2. Wilmington, Del.: Scholarly Resources.

Schmink, M., and C. H. Wood. 1992. *Contested frontiers in Amazonia.* New York: Columbia University Press.

Turner, B. L., II, J. Geoghegan, and D. R. Foster. 2004. *Integrated land-change science and tropical deforestation in the southern Yucatán: Final frontiers.* Oxford: Oxford University Press.

Vandermeer, J., and I. Perfecto. 2005. *Breakfast of biodiversity: The truth about rainforest destruction.* 2nd ed. Oakland, CA: Institute for Food and Development Policy.